Securing the Network:

F. Scott Yeager and the Rise of the Commercial Internet

ISBN: 9781520155586
ASIN: B01N4CQX1F

Revisions & Updates:

The first edition of this book was released December 31, 2016. Slightly over two weeks later an updated version with additional grammatical editing and a new, professional cover was released. This update follows three weeks later with additional minor corrections and updates. Differences from the original are minor and cosmetic, plus some minor factual errors are corrected thanks to feedback from supportive early readers. Additional corrections are welcome.

Documentation and Video Links

This book includes references to vast amounts of supporting documentation, videos and reference materials. We invite the reader to explore the additional materials, documents and videos provided on: http://fscottyeager.com/

A vast array of supporting documentation is provided on the companion website. Many of the documents are PDF files and some browsers have difficulties opening some older PDFs. If you have difficulty with any links, check your browser's PDF reader.

We make extensive use of hyperlinks and QR codes to enable easy access to these supporting materials. Any Smartphone or Tablet can use a free QR Scanner app, and simply scan the graphic for immediate access to the video or document. The QR Code image itself is also a hyper-link and may be tapped to follow the link. Obviously, this only works on the eBook version, whereas the QR Code works optically with any format, including the printed page. QR Codes are used for access to supporting website materials. Links and footnotes connect to other reference sites, such as Wikipedia. Human-readable URLs accompany the QR codes as footnotes.

We include lengthy quotes from Scott's earlier autobiographical notes. For the printed version, we have adopted a convention of using a symbol of three dashes inside brackets [---] as an eye-catching flag for the change of speaker. This is to aid the paperback book reader. In the eBook, the quoted text is in a blue color for differentiation, but the paperback is limited to monochrome.

About the Author

Nathan Gregory has 30+ years of experience in telecommunications and Internet technologies. He is a technology generalist with experience covering both hardware and software disciplines in many related fields. His present day-job is in Internet Security, where he serves as CTO and Chief Scientist for a novel Internet company named Reprivata.

Nathan's writing includes a series of Science Fiction novels and a massive genealogical research effort documenting the origins and mythology of the Gregory families in America.

Although a niche market, it is of interest to anyone named Gregory or married to a Gregory, or descended from a Gregory. It is available as a PDF on Google Books, easily found by searching *'Gregory Family Origins'*. A second edition with significant updates is planned for release in 2017. If you have any insight into Gregory history, he would love to hear from you. Please contact him via his Author's web pages.

Acknowledgements

The original source material in this book comes from extensive notes and documentation supplied by Scott Yeager, much of it from his earlier efforts at an Autobiography. I volunteered as a favor to my long-time friend to pull the previously existing work together, add my own memories and notes, and incorporate the many conversations and other documentation provided by other parties present for the events described. I received no pay for this effort, nor do I seek author credit for anything other than the minor additions and personal memories I contributed.

For much of this book I am merely a scribe and copy-editor, doing my best to make this historical tale accurate and readable. My only purpose is to document the early history of the Commercial Internet, a story largely untold before now. For, although this is a biography of Scott, it is also a larger history of how the Internet as we know it today came into being, the role Scott and I played, and give due credit to many others who were involved and made their own significant contributions.

Readers who were there with us are invited to send their own notes and experiences for inclusion in later revisions or possible future works. We welcome feedback and constructive criticism from anyone who was there and participating in the historical events described.

This is not an ego-driven tale of a "one man band" that single-handedly built something, but instead a humble recounting of a monumental effort by many smart, capable and dedicated people who worked to implement a shared vision of what the Internet may become. An index in the rear of the book lists many of those people to give them appropriate credit.

Foreword

I met Scott Yeager in 1992 shortly after I left BT Tymnet along with my friends Al Fenn, Ken Holcomb, Bill Euske, BJ Chang, Ron Whitlock, Dan Lasater, and Steve Feldman to join a networking startup that became influential in Internet history. The events that followed — the creation of MFS Datanet; the conception of the Commercial Internet; and the championing of the concept by Scott against immense opposition from those who failed to grasp the vision — have changed our world forever.

I was asked to write this biography of my friend as we have a shared history. Thanks to his archive of source materials and copious notes, the task has been easy. The reader is invited to explore his additional materials, documents and videos provided on http://fscottyeager.com/

A significant portion of the narrative herein is a transcription of notes and documents saved by and commentary provided by Scott, with my contribution of editing for grammar, readability and formatting. I include the tale of my own involvement in many events, and my own viewpoint.

This is much more than a biography of one man. It is the history of how the government-funded NSF operated Internet of the 1980s became the media-rich, business enabling, world-wide Commercial Internet we use today, a transformation that seemed most unlikely at the time. — Nathan Gregory, December 2016

Table of Contents

Out of Chaos …

In 1666, the City of London suffered one of the great disasters of Civilization, today known simply as "The Great Fire." From chaos came greatness. London emerged stronger and better, with beautiful architecture and great landmarks. The masterpiece of that rebuilding was St. Paul's Cathedral, completed in 1710.

The architect of that rebuilding was named Christopher Wren.[1] Upon his passing, his epitaph concludes with the Latin phrase *Si monumentum requiris, circumspice*— "*If his monument you seek, look around.*"

From the chaos of the early DARPA, ARPANET and NSF-funded NSFNET has emerged a globe-spanning communications facility we today call simply "The Internet." It has become so commonplace and so taken for granted that Wired News has decreed that writers should no longer capitalize it, although that position has sparked[2] a rebellion as others argue[3] that it is indeed a proper noun.

Like the London rebuilding, armies of resourceful and talented people, from lowly tradesmen to visionary architects, have devoted their lives to the project. It is impossible, not to mention absurd, to single out any individual as having "invented" the Internet. Nonetheless, just as Sir Christopher Wren was an architect and visionary leader in the London efforts, certain people today stand out as leaders who made monumental contributions to today's Commercial Internet. In this work, I choose to capitalize both the word Commercial and Internet and treat it as a proper noun, because they refer to a specific thing derived from the original Internet and in which I have been privileged to participate as a lowly tradesman.

We all know of the visionaries who made their contributions. The list is long and hosts many prestigious names. Marc Andreesen and Eric Bina developed the first Internet Browser; Steve Wolff managed the NSFNet in the 1980s; David Clark was the chief architect of many of the protocols and the formation of the rules governing the Internet; Jon Postel and Paul Mockapetris designed and developed the global DNS system; Vint Cerf is a legend who wore many hats, Larry Roberts led the development of Arpanet and founded Telenet, which we cover in more detail later; Douglas Englebart is another legend, known for inventing the "Mouse" and who wore many hats, including that of Fellow at Tymnet where I worked in the 1980s and was privileged to meet him. The list is virtually endless; I could easily list another one hundred names in this pantheon.

This book is not about any of the well-known Internet luminaries, noteworthy though they may be, nor it is an epitaph. It is, however, a biography of someone who should receive credit for his role in turning the chaos of the early Internet into the media-rich digital medium we casually take for granted today. It documents precisely his role in that regard and his true accomplishments. In a larger sense, it is a biography of today's media-rich Commercial Internet.

---

[1] https://en.wikipedia.org/wiki/Christopher_Wren
[2] https://www.wired.com/2015/10/should-you-be-capitalizing-the-word-internet/
[3] https://en.wikipedia.org/wiki/Capitalization_of_"Internet"

In the 1980s Scott Yeager was an early driver in the move toward carrying LAN traffic beyond the office walls. In the 1990s he was an innovator who envisioned the interconnection of disparate networks into "peering points" and who wrote the rules that allowed the explosive growth that followed. In the 2000s he created the streaming media we use today.

Scott Yeager was the visionary driver for the formation of MAE-East, the long-haul backbone at MFS Datanet, and a forward-thinking pioneer of the media-rich streaming data services that came into being at the ill-fated *Enron Broadband Services* (EBS) prior to the dot-com bubble burst in March of 2000.

The time has come to tell the other story of the Internet, the one that today is almost unknown, the role of Scott, myself and many others in building today's Commercial Internet, and share the vision Scott and I hold for what it must become. This book provides not only the story but backs the tale with immense volumes of supporting information, documentation, videos and proposes a future that we believe must logically follow.

This tale is not singularly focused on the past. It tells not only how we got here, but where we think the Commercial Internet must go. For all its greatness, today's Internet has serious shortcomings. Theft of personal data, identity theft, online scams, and advertising fraud run rampant, with online dollars diverted to organized crime. Insecure systems, poor security practices and an attitude of secrecy and reluctance to acknowledge failings inhibit real solutions. We propose a way forward, a networking future that is bright, optimistic, and secure, and have created a company called Reprivata to guide it.

Many people *think* they know the Enron[4] story when they recall press accounts of faulty accounting practices and executives being trotted off to prison. Though indicted, Scott was cleared of all criminal charges. When the government attempted to try him a second time on essentially the same grounds, he battled this second case all the way to the U.S. Supreme Court. And, again, he prevailed.

The scandals are what people remember. But there is another story, a less well known, seldom discussed side of Enron that emerged under the name of *Enron Broadband Services* (EBS) in 2000 on the eve of the dotcom bubble burst. Though largely forgotten today — EBS barely gets a mention in the Enron profile on Wikipedia, for example — Wall Street and high-tech firms in 1999 and 2000 recognized EBS as a major player in the technology sector.

There was a stir of genuine and warranted excitement in the air. Dotcoms were eager to enter into business partnerships with EBS because EBS was going to enable e-commerce, large scale file sharing, video streaming, and application services flowing through a distributed server system and fiber optic network configured on a scale never before imagined. This was rooted in work Scott had done at MFS and which WorldCom had failed to embrace. Enron hired him specifically to build what WorldCom had declined.

---

[4] https://en.wikipedia.org/wiki/Enron

The basic concept was known as a "distributed server application service" and with it EBS was creating a *Content Delivery Network* (CDN).[5] These application layer services were called the *Enron Intelligent Network* (EIN) and it ran on the *Broadband Operating System* (BOS). The CDN over which EBS would deliver streaming video was called Media Cast and Media Transport. These EBS innovations followed from earlier innovation at MFS, and that had followed even earlier innovation as a solitary entrepreneur.

The story of Enron became lost in the scandals and false claims of the prosecutors. The claim of prosecutors was that the BOS and EIN technology at EBS did not exist, wasn't real or didn't really work. Nothing could be further from the truth.

The technology, innovations, and the future envisioned at EBS was real, and it's all around us today. Today's Commercial Internet grew directly from those innovations. The Commercial Internet includes media-rich streaming content to enable e-commerce, and engage large audiences via all types of devices. This is the story of the commercialization of the Internet, but also about its flaws and how it now enables bad actors to take advantage of users, as well as what we can do about it. *Circumspice!* – Look Around!

---

[5] https://en.wikipedia.org/wiki/Content_delivery_network

FROM STUDENT TO ENTREPRENEUR

(1975-1981)

If somebody were to ask me how Scott found his way to Enron, one quick and simple answer is that Enron found its way to him. Enron moved its corporate headquarters from Omaha, Nebraska to Houston.

Another answer is this book. By which I mean, all of what follows, beginning with graduation from the *University of Texas at Austin* with a *Bachelor of Science* (B.S.) in Biology in 1974. He had worked hard to get his degree and was done with school. He had no desire to get a Ph.D. or Masters. Scott just wanted to get on with life and put to practical use the theoretical studies of the years in school. And, no question about it, those studies would prove deeply helpful. Above all else, it instilled a knack for problem solving.

While in college, his electives consisted of two semesters of calculus and differential equations, two semesters of physics, and four semesters of chemistry (including two semesters of organic chemistry). These courses all taught disciplined problem solving. Setting up the problem, analyzing the knowns, the unknowns, and the variables was the interesting part. The solution required deriving the proper equation addressing those factors. Building the optimal equation was the challenge, the interesting process. The mundane math to derive the solution was far less challenging than developing the correct equation. We both welcomed the arrival of computers and spreadsheets for the mechanization of routine calculations. We both were early adopters of these tools.

Like my own case, the appreciation of the scientific method came long before college. In my case, that appreciation came from a High School mentor, a physics teacher we all called "Doc." In Scott's case, St. Thomas High School in Houston, Texas, similarly instilled a solid foundation in science and math.

The Scientific Method has characterized natural science since the 17th century. This codified process of systematic observation, measurement and experiment, and the formulation, testing, falsification and modification of hypothesis is the one true pathway to knowledge and the dispelling of erroneous beliefs. I hold the opinion that the discovery and codification of the scientific method must rank as perhaps mankind's greatest achievement, for so very much else we have accomplished flows from that process.

Problem-solving and the discipline of the scientific method are essential tools in developing an approach to innovation. These are the tools used to find new ways to solve problems and create new approaches to common problems. This elementary understanding has proven useful in guiding both of us throughout our lives.

Science and the Environment

After college Scott went to work for an environmental consultant, Dr. Nugent Myric, Ph.D. Dr. Myric assigned him a project developing an environmental impact statement for the *Brazoria County Airport*. Scott's college roommate, Mike McCullough, also worked on the project. They determined what was required by studying environmental impact statements for other airports. It was a heady experience, fresh out of college, to develop the justification for building an airport.

Then came the challenge. A pair of bald eagles were nesting within a mile of the runway! Could they prove the eagles would not be impacted by the aircraft?

Scott wrote the *Environmental Impact Statement* covering all the concepts and parts that needed covering. Except for the eagles. Less than two weeks from the due date, he admitted to Dr. Myric that lack. He planned to go to the Rice University Library and research the issue until resolved. Today we would merely tap Google, but in 1974 "search engine" was a null term and the Library card-catalogue[6] ruled.

Those pushing the project subtly implied the answer had better be favorable. Scott was troubled by the pressure to show the outcome in a favorable light and yet adhere to the principles of science and perform good, unimpeachable scientific work. This is often a conundrum in scientific research, adhering to the science in the face of pressure for a specific outcome. This early experience taught Scott valuable lessons in the need to understand the political landscape and the rules the other side plays by. Scientific research is so very often about much more than science.

Scott went to the Rice University library as planned and spent nine days searching for prior work that might shed light on the question of environmental impact on the eagles. He finally found a decade-long study from Alaska. That study sought to learn how many eggs eagles typically laid and how many hatched.

The researchers faced the problem of how to peer into and study an eagle's nest, typically located in the highest spot around. Deep in the body of their study was a reference to the use of small fixed-wing aircraft to attempt to scare a mating pair off the nest. To observe the number of eggs in the nest they thought they could scare the parents off by buzzing aircraft around them.

The tactic failed. It did not cause the birds to move. Nor did it change their mating habits. The study, in fact, found no impact after thousands of attempts at buzzing eagles' nests over a 10-year period. Intentional harassment by aircraft did not bother eagle nesting habits at all!

The logical conclusion followed that an airport one mile from an eagle's nest would pose no threat. One sentence in the study said it all. "Small fixed wing aircraft had no negative impact on the mating habits of bald eagles."

Scott's work was reviewed by key environmental players, including the Audubon Society, and met with general acceptance. It also sailed through the official public reviews and approvals with no problem.

It was Scott's first success as a member of the workforce, and a formative one. He found it gratifying and confidence-building to know he could find the answers by performing careful research and following the science.

---

[6] https://en.wikipedia.org/wiki/Library_catalog

Science to Business

Bruce Hotze, a childhood friend from grade school through college, helped to start a family business manufacturing piston-type compressor parts in the family garage while Scott was in high school. Scott worked for his friend's business in its early days for $1.60 an hour, the minimum wage between 1968 and 1974. That's roughly equivalent to $11 per hour today, good pay for a high School kid.

After college, Bruce asked Scott to come to work full-time for *Compressor Engineering Corporation* (CECO)[7] and leave his environmental consulting job. He wanted Scott to start an inside sales group and help with other things. By 1976 it had become a respected small business with 25 employees. The offices were in an old house and there was a 2500-square-foot manufacturing facility across the street.

Scott soon discovered the practical application of technology in everyday life employs the scientific method. Observation of what customers said they wanted is only the first step in a chain of problem-solving that ultimately determines what they need. Then, the test of the hypothesis comes in determining what they will pay for. Falsification comes when these things do not match and a new hypothesis emerges closer to the intersection of what the customers say they want, what they need, and what they will buy.

The time at CECO was especially valuable to Scott. Several experiences there set his life course. This is where he learned about computers and their need to communicate over cables. This is where he learned to build and manage a sales organization, and this is where he learned how the phone company worked.

He liked that CECO was small and that they were to figure out how to build sales and expand the number of products. Bruce had more faith in Scott's abilities as a salesman than Scott himself did at the time. He had not considered being a salesman after college, although he had sold aluminum siding and roofs in underserved neighborhoods his junior year.

Another mutual friend joined the same day. Bill Zeis attended grade school, high school and college with Bruce and Scott. Bill had received a journalism degree from UT Austin and was a great writer and very smart.

The kitchen in the old house serving as CECO's business office was designated their workspace. The counter and sink were torn out and replaced with used metal desks. Scott's desk nudged against pipe stubs that had been connected to the missing sink. His executive view was pipe stubs and patched, unpainted sheetrock.

There was no air-conditioning in the room, although there was a window unit in the living room where Bruce and his secretary had their offices. They bought a fan and redirected some of the cooler air into the kitchen. It was basic and functional. They were building a company from scratch without wasting money on trivia.

Bill's desk was next to Scott's. They were officially employees. They looked askance at each other and laughed. Neither knew a whit about business or sales. Still, they shared the sense that they could help make something out of CECO, even if they had no idea what they were doing. They knew they were in for a new experience.

---

[7] http://www.ceconet.com/

Then Came the Computer

In early 1977, Bruce and his brother Jim, who was the CPA in the family, decided to "computerize" the company. *Digital Equipment Corporation* (DEC) had released the *PDP-11/10* minicomputer.[8] CECO purchased a system to create an accounting system for the company. That is the same year Apple Computer, Inc. was incorporated and the Apple II was introduced in April. August of that same year would see the introduction of the TRS-80, which I would soon buy to begin my own digital journey. These were virtually toys, Scott got to play with a real computer. Yes, I am jealous.

Scott and Bill were tasked to 'figure it out' and cover the basics in documenting what was going on in the business. They were to learn how things were being done first on paper and then to give that feedback to the programmer. This was heady stuff, as educational as any college course.

Not only were they ignorant about computers, neither even knew what a purchase order was nor an invoice. They had no knowledge of inventory control systems, manufacturing processes and had never seen an industrial compressor. CECO made reciprocating compressor parts, a piston-driven compressor versus centrifugal compressors which used a spinning impeller on a shaft to compress the air or gas. They didn't know the difference until Bruce explained it their first day.

In fact, Scott says neither of them knew anything about what the business did. Despite their ignorance, they were being asked to learn the company's inner-workings. There could be big ramifications for the entire company. They made sure Bruce and Jim knew how clueless they felt. "Go figure it out," they said. Uncertain but confident, they did just that.

First, they must learn what the company did. That started with the shipping department under the tutelage of a lady named Yolanda. She explained how orders arrived from a salesman. They wrote it up on an order form, using five-page carbon paper and had to write very hard to make all five copies come out legibly. Then one copy (page three of five) went to Yolanda and she would pull the parts from inventory. Each item had a part number and often consisted of assemblies of smaller parts. For example, there was a part number for a "Channel and Spring Set" made up of six, eight, 10, or 12 channels of differing lengths and the matching springs.

She explained that this was a product that CECO had become known for. They made the best channel and spring sets for Ingersoll Rand compressors anywhere, and they were in stock. Both were a big deal to the industrial plants and compressor pipelines pumping natural gas across the USA.

She showed them how she pulled the separate parts from inventory to make the set for the part number and then recorded all the component parts and how many she pulled. She did the math and verified how many were left. If one of the separate parts were out of stock she could not make up the combined assembly part number so it was put on hold until she could obtain all the parts for the combined assembly.

---

[8] https://en.wikipedia.org/wiki/PDP-11

It was all manual and when she went to a warehouse bin and pulled the parts, pointed out the part number on the individual part and how she matched it to the assembly numbers, the process began to make sense.

They continued learning about the inner workings of CECO. They learned that the manufacturing workers made the individual channels and springs. They learned those parts had to be sent out for "rolling the channels," the process of heat treating the channels and springs. Then they were placed into inventory after they were tested and approved.

There were engineering drawings for each part that were "blue lines" – an ammonia process that made a copy from the original pencil drawing created at a drafting table by an engineer and a draftsman.

He learned what a "blueprint"[9] was.

This all had to come together for each part number, the individual parts, and the assembly process. That was called a "Bill of Materials" or BoM. They wanted the computer system to track all this information, make it available to the different departments and have each department pull key information and update it as things happened. The chain started with the engineering drawings, went to manufacturing, which included purchase orders for the vendors and internal documents for the internal manufacturing process, then to inventory or the shipping department if it was not an inventory item but "made to order." Then it was put in inventory until some customer ordered the product. It was then written up as an order, pulled from stock and shipped. Then it had to be invoiced, paid by the customer, put in the books and tracked to show a profit per line item.

"Computerizing" the Workflow Process

The computer had been purchased for accounting, but tracking workflow processes naturally followed. As Scott's understanding of the business grew, he saw where automation could be applied.

Each location where someone needed to look things up or enter information needed a computer terminal. They had to run cables from those locations inside the production facility back to the DEC PDP-11/10 computer in the computer room in the front of the building. This was a formative experience and a genesis of a future business path Scott would follow. This project became perhaps one of the most formative experiences of his life.

Inventory control and shipping were in the back and the computer was in the front. The accounting was close to the computer so it was not difficult in this small manufacturing building. There were very few users early on and getting a terminal to use was costly and a big deal so not many people got to access the computer system. Not at first, anyway.

It took about six months to figure everything out. They spent a lot of time with a consultant, plus Bruce and Jim of CECO discussing the issues of how to best "computerize" the company.

---

[9] https://en.wikipedia.org/wiki/Blueprint

Selling via Long-Distance Telephone

Before deregulation and AT&T provided all telephony services through their regional divisions. *South Western Bell Telephone Company* (SWBT), then a division of AT&T, was the local telephone monopoly. It did not become a stand-alone company until "spun-off" by a January 8, 1982, court decision.[10] They sold both the local "dial tone" service and long-distance minutes. Long-distance calls cost about 75 cents per minute in 1976. For perspective, a three-minute call cost about the same as one hour for a minimum-wage worker. Long-distance was costly.

SWBT sent a sales representative to talk about new ideas. Bruce asked Scott to sit in and learn. The representative explained how long-distance could save the business money. It was cheaper to make sales calls over the phone than it was to send a sales representative to call on a customer.

Another sales approach was to send out a "form letter" that stated specific things and the computer system the added information that was unique to the targeted customer. Scott saw this was a natural application for the computer system. Scott explored the benefits of writing the "form letters," and developing the custom inserts so each letter looked like it was hand-typed, although it was mostly from the computer. This was a radical idea.

Inside Sales Team

Bruce asked Scott to hire and develop the "inside sales team" to make calls and cultivate new customers. Again, with zero experience at any of this and no sales experience. The goal was to land prospects and make phone calls to customers who bought rarely vs. a customer who ordered regularly.

CECO soon moved to a larger building. This provided room to expand the sales department and grow the business. They were doing well and he had great authority and input into the growth of the business.

Scott's "inside sales team" gained traction quickly. They were good with customers and effective. They performed well and soon accounted for almost 50-percent of the total sales, vs. the "outside sales force." They also lacked the travel costs, although the long-distance costs were significant. The computer reports showed the cost of sales for his group was a fraction of the cost of sales for the outside sales force. Bruce was teaching Scott all aspects of building a business including tracking costs of good manufactured and cost of goods sold as part of total overhead to arrive at how much profit the company earned per item.

Soon Scott managed up to ten sales representatives. They were selling nation-wide using long-distance calls. He trained his team to use the computer system to provide answers to customers faster than ever possible before.

---

[10] https://en.wikipedia.org/wiki/United_States_v._AT%26T_Co.

International Sales Group

The next stage became creating an international sales group. The company expanded and hired a manager for inside sales as they began selling compressor parts all over the world. They had the ability to quote big projects by then and were using the computer system more and more to get quotes out the door and keep track of the business. The computer system constantly needed improvement.

Bill was now responsible for purchasing and used the computer as one of his key tools. Scott represented the view of sales, Bill represented purchasing, Betty represented manufacturing, Art oversaw engineering, and Yolanda oversaw shipping. It was a growing system that covered the entire company. Jim used the computer data to generate the profit/loss reports as the CFO and Bruce used all the information. reports and input from everyone to run the company as the CEO.

Scott and Bill accomplished a lot since joining the company. They had adopted a telephone sales approach, created a computer system to automate the business and provide support to the telephone sales team, produced computerized "form letters" that customized quotations for each customer.

Scott developed an international sales organization that used the computer system to quote products, book orders, and get products manufactured and shipped, invoiced and paid. He was unofficially directing "product development" because many customers would place special orders large enough to justify new products they'd never made before.

Scott's time at CECO was a confidence instilling experience. The business education and sales experience he gained in this time was more significant than anything he'd learned in a classroom. Scott felt confident to tackle large projects and do things he'd never done before while learning from customers. The faith that Bruce had put in him and Bill had paid off and Scott was confident because he had accomplished things that seemed a huge stretch for a Biology and Journalist major with no business experience.

Now, not yet thirty years old, he was ready to take the next step and establish his own business.

BIRTH of Yeager, Sumners and Associates

(1981-1984)

Scott started his own company, *Yeager, Sumners and Associates* (YSA). It was a manufactures' representative business for wire and cable used in computer, telephone, and other low voltage signaling systems.

He wanted to "computerize" the company but the IBM PC was not yet ready for the task, and the DEC systems were too expensive for the small entrepreneur. A refurbished Wang word-processor seemed the ticket. It used a specific type of cable, dual Wang RG 59 coax cable. YSA sold millions of feet of that cable to customers as well. That illustrates how many of those systems were in service. *Brown and Root*,[11] a Texas-based multinational engineering and construction firm, used them for all their reports.

Imagine word processing specialists who typed everything. No executive, engineer or sales representative typed his or her own correspondence. Secretaries typed everything. In the past, Scott would handwrite a letter or a memo on a yellow tablet and then the secretary would type the draft and then they would edit and retype it five times to get it correct. The first draft is always very bad and it would take at least three versions to make it something professional. The development of word processors on the PC saved businessmen and their secretaries immense time and resources. They no longer had to rely on a typewriter to create correspondence.

Using a DEC computer system and dealing with Wang word processors provided tremendous insight about why people wanted to "computerize" their companies. When the IBM PC came out, Scott recognized we were at the beginning of a new trend. From 1981 through 1983, Scott believed the computer trend was just at the beginning and not a fad that might end. It seemed that YSA could be a player in the newly evolving communications and low voltage signaling industry.

Competition for long-distance minutes and the beginning of customer-owned *Private Branch Exchange* (PBX)[12] systems were major trends also developing in this time-frame. He understood that YSA was in a fast evolving industry that was going to keep growing for years.

There was no such thing as viral growth or the Internet. Small companies that worked hard could develop a name for themselves. The economy was on the upswing, and the business landscape seemed promising. YSA intended to take advantage of growing business opportunities in a fast-growing industry.

Pistons to Cables

Scott left CECO in 1981 to found his own company, choosing the wire and cable business with a focus on voice and data cables as well as other low voltage systems. It was a scary change. He feared it was crazy to move from an industry he understood and was doing very well in, to a new business he knew nothing about and must learn about from the customers.

---

[11] https://en.wikipedia.org/wiki/KBR_(company)

[12] https://en.wikipedia.org/wiki/Business_telephone_system - Private_branch_exchange

He was now an independent manufactures representative for a wire and cable manufacturer. This meant he was a pure commission sales representative, but could build a business representing more than one manufacturer. The first cable manufacturer YSA represented was Berk-Tek out of Reading, Pennsylvania. Scott developed a good relationship with their owner, sales management, product development and engineering people. These personal connections became essential in influencing the factory to develop new technologies and help meet changing customer requirements.

Motivating the factory to make products for customer's needs was a survival skill that was essential to success when selling on a commission-only basis. Orders equaled revenues. And if he did not get orders he did not make anything. This put pressure on Scott's wife Susan and the kids since they needed the money to live and he had to travel to get orders in addition to using the phone to spread himself around more and cover more territory. It was a tough business with every cable being treated like a commodity, and any value-added service business was almost non-existent.

YSA sold plenum rated cables for energy management, fire alarm, telephone and computer systems. This was a new kind of cable created from a change in the *National Electric Code* (NEC)[13] that mandated the use of fluoropolymers that would not burn and not propagate flames in the air space between floors in a high-rise building in the event of a fire. DuPont made Teflon as an insulator and jacket material but it was expensive. Pennwalt made Kynar and it was another fluoropolymer that was more brittle but a lot cheaper per foot. YSA sold a lot of that. Berk-Tek pioneered methodology to extrude Kynar onto wire so they led the way in plenum-rated cables.

He discovered that the best way to sell this cable was to find major construction jobs or major retrofits of computer, telephone, energy management or fire alarm systems. These large projects would enable YSA to bid on and sell the cable to the electrical contractors.

He had to sell the architects to get the concept allowed in the specifications and then sell the contractors so they would build it using the Berk-Tek plenum cable instead of other brands. He approached the architectural firms and educated them on plenum vs. PVC cabling. PVC can give off toxic fumes if it burns, plenum has a flame-retardant coating. He persuaded them to approve plenum cables in lieu of using conduit combined with PVC-coated cables. Conduit and PVC coated cable was the standard for architectural firms. Scott would go to the contractor and explain the alternative of plenum rated vs. PVC cabling and convince them to choose Berk-Tek plenum cable over using conduit to win the bid. They would do it because it was a much less expensive installation choice.

Another problem was distribution. YSA also had to line up distributors and convince them to put the cable in stock so the contractors could get it off the shelf. Convincing customers to buy, if they couldn't get the product, was futile. Scott was successful at getting Wholesale Electric in Houston and Wallace Electronics in Dallas to carry the cable in stock.

---

[13] https://en.wikipedia.org/wiki/National_Electrical_Code

It was difficult to create a new market for plenum rated cables from scratch. Especially since he was recently from the reciprocating compressor parts business and lacked experience selling cable to contractors. But he did know how to sell technical products that made a difference to a customer. Scott applied that knowledge and experience gained at CECO to selling plenum rated cables to owners of buildings, tenants in buildings installing systems, architectural firms, and electrical contractors.

Scott describes the experience like this:

[---]

I had been a *"telephone man"* and installed cables in high-rise buildings while attending UT. We installed them in a 'star' design, which meant bringing them back to the main phone room where the PBX equipment was located inside the tenant's building. This was a different cabling architecture than installing cable inside a home or an apartment that ran a loop through all the locations and 'dropped' different pairs of the cable for the different rooms or apartments.

This background meant that I understood cabling systems even though I had never been formally trained in them. I had learned that home phones were looped through the house with the same pairs of wires showing up in all the rooms in the house. I also learned that business phones like PBX telephone exchange cables were laid out just like data cables for computer systems. They were the same "star" architecture where you made a 'home run' back to the computer main system in one room. This was terminated in a patch panel in that room to allow each desktop to have a computer terminal connected to the main computer. At that time, the main computer was either a mainframe or a mini-computer. In 1973-1975, mini-mini-computers and PCs did not exist. In the early 1980s, that telephone cabling experience helped me immensely in understanding the installation of computer cables.

Between 1981 and 1983 I took the sales of plenum cable from zero dollars to several million dollars in revenues per year. Berk-Tek became a key product line for low voltage cables in Texas and surrounding states. I was highly motivated, of course. I had to make sales or I'd be unable to cover my bills. It was not easy working on a pure commission sales basis and having to garner orders new orders every day, week, and month – or else go broke.

First, you had to get enough orders to pay the travel expenses, office rent and long-distance bills. The single largest expense item was long-distance calls to customers. I racked up more than $2,000 each month in phone bills and then spent another $3,000 each month in travel expenses. Luckily, Southwest Airlines was in business then and I could fly to Dallas, Austin, San Antonio for about $50 for a one-way ticket. I'd just go to the airport and jump on a plane. It was a huge benefit, money and time-wise, to be able to fly to Dallas in the morning and return in the evening and not pay for a hotel.

Later when my wife's brother, Philip Sumners, became my partner we had to pay for two families out of sales commissions. It was not easy to get up every day and go get orders for telephone, energy management, fire alarm, or computer cables. You sure did learn to work hard and not take anything for granted when you did succeed in getting orders. The intensity of always being about to go broke made you very keen on how you spend your time, how you followed up on deals and who you spent time on in the sales process.

The biggest computer company in the world was IBM which had started making electric typewriters and evolved into the world's largest multi-national computer company. IBM was so big that the *Department of Justice* (DOJ) tried to break them up in a huge multi-year monopoly-busting lawsuit. DOJ was also working on breaking up AT&T's monopoly of telephone services.

[---]

The antitrust laws are a uniquely American institution. They came into being in the 1890s with the Sherman Act, which prohibits the exercise of monopoly power. That's when a company that dominates its industry uses its power to keep out competitors. Antitrust laws pit the interests of small companies against those of big ones. And somewhere in there is the desire to protect consumers by giving them competitive markets, which are supposed to push down prices and boost quality.

The DOJ's attack on IBM began in 1969 and was finally dismissed in 1982, only to be attempted again in 2009, though soon dismissed.

Followers of the Adam Smith *"invisible hand"*[14] argue that monopolies endure because of government meddling in the marketplace, not despite it, and that government attempts to break up monopolies in the name of fairness are misguided and often don't work. The idea asserts that free markets and open competition automatically destroys monopolies. Certainly, the telephony market, with the long-held "protected monopoly" of AT&T illustrates how Government-fostered distortion of the marketplace can perpetuate monopolies.

Government distortion often means that favoritism is given to those with the deepest pockets. Today we see as much monopoly behavior by today's reincarnated AT&T as when the old AT&T was a protected monopoly, owing to lobbyists and other influences. The same goes for countless other companies. Government monopoly-busting has a disappointing history, and free-market competition, when it arises, is a far preferable mechanism to dismantle such monopolies. The Telecom Act of 1996, for example, dealt the would-be monopolies a serious blow as competitors arose to do battle. It took years for them to reassert their control. Sadly, the forces that fostered the Telecom Act were absent when the would-be monopolists regained their position. Today, much of the web-surfing public effectively has no choice of service provider.

---

[14] https://en.wikipedia.org/wiki/Invisible_hand

LAUGH-IN to NASA

There were several movies about the phone company controlling everything. The movie *The President's Analyst* (1967) was a spy thriller in which the guys controlling everything are discovered to be "The Phone Company" (TPC). It was a common feeling in the late 1960s when AT&T was a monopoly. The phone company was indeed everywhere and extraordinarily powerful.

At the time, a talented comedian named Lilly Tomlin was also giving voice to growing consumer concerns. Tomlin did an ongoing bit about the phone company on the wildly popular TV show Laugh- In (1968-73). Tomlin played an old-time switchboard operator who worked with a manual telephone switchboard. A live operator connected your calls for you, and callers had to tell the operator who they wanted to be connected to and in what city. This could lead to friendly and personal exchanges like those enacted by Tomlin in her skits. But by the 1970s the old technology was being phased-out, and there were huge electro-mechanical switches in the big central office buildings spread around town. Tomlin's character poked fun at the phone company for using old-timey technology in which living people manually inserted wires, switched calls, and were free to intrude on people's conversations in funny and snarky ways.

Not only would her character, Ernestine, "horn-in" on customer conversations, she would hilariously mock the callers, and did not at all take the customer or her job very seriously. Her famous rejoinders to customers who objected to her obnoxious interjections, include: "'Fair?' Sir, we don't have to be fair. We're the phone company."

Both the movie and Tomlin's *Laugh-In* skits hit home, giving voice to a dissatisfaction many people were feeling. Many people did not like the phone company and did not trust it. The slow pace of innovation and insular service was evident even before the skits and movie.

The image problem the phone industry bore foreshadowed and contributed to the changes that would come later. Of course, the wide scale adoption of computer technology is another phenomenon that signaled the coming change.

During the 1960s and 1970s whenever mainframe computers were mentioned, it was usually IBM that came first to mind. There were others, but IBM was the "big dog" in this space. These were computers that required housing in a huge computer room.

Mainframes had tremendous processing power for the time and took up a great deal of space. To put a scale to the power of these systems, a single 1977 vintage IBM 3033 Mainframe[15] with a mere 16 MB of memory could in theory support over **Seventeen Thousand** remote terminals under CICS, IBM's operating system.

That represents a lot of cables. IBM mainframes were state-of-the-art computing, extremely expensive. All the big financial institutions, oil companies and government entities had them. IBM was the undisputed leader in mainframe development and sales.

---

[15] https://www-03.ibm.com/ibm/history/exhibits/3033/3033_PR01.html

The user connected via a terminal, (a 2260/2848 in the 1960s, or a 3277 after about 1971), which had a green screen with a keyboard. The IBM terminal used RG62 coaxial cable that was "home run" back to the main computer room and connected using BNC connectors. YSA sold millions of feet of both the PVC and plenum rated versions of this RG62 cable for all kinds of IBM computer systems spread all over Texas, Oklahoma, Louisiana, and Arkansas.

It was in this way that Scott became aware of how big the computer cabling business was and recognized there was a boom underway where all larger businesses were installing computer systems.

The science gurus at NASA and research departments at the oil companies started buying less expensive *Digital Equipment Corporation* (DEC) computers. They used terminals as well. When he had worked at CECO they installed DEC VAX computer systems and ran cables to those terminals. This was a mini-computer, small compared to an IBM mainframe system. This experience helped him to understand the evolution occurring in the early 1980s. Computers were being installed everywhere and would soon become central to most businesses.

DEC implemented an idea conceived by Bob Metcalf at Xerox's *Palo Alto Research Center* (PARC) in 1973-1974. Xerox was then known only for their copy machines. This innovation was called "Ethernet" and Xerox patented it in 1975. It allowed the creation of *Local Area Networks* (LANs). The first implementation, in 1975, was a 3 Mbps connection for the Xerox Alto.[16] Bob Metcalf left PARC in 1979 to found 3COM and introduced Ethernet commercially in 1980. He convinced DEC to use it.

It became standardized as IEEE 802.3 on June 23, 1983. In the early versions, a special coaxial cable (A doubly shielded variant of RG-8/U, 50-ohm coax came to be called "Thicknet,"[17] a later variant used RG-58/U became known as "Thinnet"[18]) was required, the Ethernet bus cable. The drops used twisted pair cable to get to and connect with DEC workstations using an *Attached User Interface* (AUI) Transceiver. This was specialty made for them by companies like Berk-Tek, which YSA represented. So, Scott sold some of the early Xerox Ethernet cable and the early DEC Ethernet cable as a representative for Berk-Tek. This is how he became aware of the growing trend away from IBM host mainframes to LANs with lower cost computer systems and software that ran on less expensive systems like the DEC computer systems.

The DEC "workstation" was unique to DEC and the IBM "terminal" was unique to IBM. This meant that the cable had to be special as well. Everything was custom made for each system. Each terminal required one unique cable per user to connect one terminal back to the IBM mainframe rooms or the DEC minicomputer rooms.

DEC was making computing low cost enough for smaller companies and IBM was starting to feel the heat of competition. IBM owned both the big Enterprise company and government market for computers. No one thought this would change; IBM was the behemoth company that ran the computer market worldwide.

---

[16] https://en.wikipedia.org/wiki/Ethernet - cite_note-Alto-3

[17] https://en.wikipedia.org/wiki/10BASE5

[18] https://en.wikipedia.org/wiki/10BASE2

Free-market competition was emerging from DEC and soon others would follow. By 1982 the DOJ would realize that competition had already resolved the monopoly concerns that had sparked that 1969 anti-trust action, and would finally dismiss the case as "without merit."

Arrival of the PC

In 1981 IBM released the first "Personal Computer," or PC, as the world has dubbed these indispensable machines. They were not the first to market, but they legitimized it in a serious way that caused people to notice and take seriously. Apple had their Apple II introduced in 1977, and Radio Shack had their TRS-80 also introduced in 1977. Various other vendors (Altos, Heath, IMSAI, MITS, NorthStar, Xerox) offered platforms built on CP/M[19] as far back as 1973. At the peak of the CP/M era, there were over a hundred different brands. These were limited machines with limited software and were not yet taken seriously by the market or casual users. They were expensive toys and only the truly dedicated users could get useful work out of them. They were far from useless, but they were difficult to use and required a degree of arcane knowledge most people found daunting.

Operating Systems were the weak point in those early computers. The lack of commonality between brands was a major hurdle. The whole Microsoft origin story in which Bill Gates acquired the rights to develop the *Operating System* (OS) for the IBM PC using DOS is a sign of how little the incumbent computer experts making the big bucks thought of the importance of software. They essentially gave away the operating system rights to Bill Gates so they could sell the hardware, which was the PC. This established MSDOS as the standard OS and killed CP/M, TRSDOS, and Apple DOS. Apple did manage to stay their own path with the Mac OS starting in 1984, but it was a battle won on sheer tenacity, fan loyalty, and technical merit.

IBM did not see PCs being used for anything other than personal use. They imagined perhaps small companies like YSA, Scott's little sales representative company, might use them as a sales tool.

Scott decided in 1981 to not buy an IBM PC because the only thing it did was run VisiCalc spreadsheets and had poor word processing software, WordStar. There was little software from Microsoft other than the DOS operating system. Microsoft Word appeared in October 1983, and Excel did not appear until September 30, 1985, when it replaced the almost useless "Multiplan" introduced as a VisiCalc alternative in 1982. The word processing software of 1981 was crude and limited. The setup required using a dot-matrix printer, the only reasonably available printer at the time, which was very crude as well. The few "letter quality" printers were glorified typewriters, expensive and cranky. IBM Selectrics typewriters could be driven by a computer, for example, but were very expensive.

WANG computer company had meanwhile developed a state-of-the-art word processing system that had a high-end printer as part of it. It looked like an IBM typewriter, which it was, and allowed you to type a form letter, add new names, and send out hundreds or thousands of personalized letters with the same content but made them look like they were typed individually.

---

[19] https://en.wikipedia.org/wiki/CP/M

However, it also allowed huge reports to be typed and edited and printed out at a drastically reduced production cost. The industry standard for how business correspondence was produced at that time was the IBM Selectric[20] typewriter, first introduced in 1961. It produced the best quality business materials. This new equipment needed cable to run the workstations. Selectrics could also connect to a computer and serve as a word-processing printer or computer terminal. The Selectric was discontinued in 1984 and marked the end of an era.

It would be years before IBM PCs had low-cost laser printers. Even dot-matrix printers were relatively expensive and there were mechanical issues to deal with so people did not use PCs for quality correspondence. High-quality inexpensive printing from a PC did not evolve until the early 1990s. Everyone owes a debt of gratitude to Hewlett-Packard, who made it inexpensive to own a quality printer and the ability to easily connect it to a PC.

The previous experience of installing a DEC computer at CECO and managing sales helped Scott to immediately recognize the computer cable market, but it also attuned him to the developing computer industry. While at CECO they got the DEC system to work for the sales department. They also used the DEC for inventory control, tracking manufacturing functions, as well as company financials. The CECO experience gave him a good, early look into what the computer industry was doing and why smaller companies might want to install computers in the workplace. Then the expression used was they were going to "computerize" the company.

Scott had learned that high-quality professional-looking correspondence using a nice typewriter was a big deal and was very helpful to sales. This was especially true if you combined that correspondence with a follow-up telephone sales call. He hired and trained an entire non-technical, non-engineering telephone sales force at CECO. They had used the combination of nice looking correspondence and telephone follow up to acquire sales all over the USA for CECO compressor parts.

[20] https://en.wikipedia.org/wiki/IBM_Selectric_typewriter

Early Adoption of Fiber Optics

In 1983, it seemed obvious that YSA must expand revenue opportunities. They made money only on sales commissions. Their biggest revenue came from copper cable extruded with plenum rated fluoropolymers. Building a long-term business meant new products. Existing manufacturers of well-established products considered YSA too small. They did not want small-potatoes as their representatives. YSA could not compete with the large, established sales representative firms. They must innovate.

YSA needed to develop sales in products that were new at that time, and fiber-optic cable was very new in 1983. Scott went to fiber optic and *Local Area Network* (LAN) trade-shows. He saw a need evolving for fiber optics in *Local Area Networks* (LANs). One of the biggest trade shows was *Fiber Optic Communications and Local Area Networks*, FOC/LAN for short.

He decided YSA should represent manufactures of fiber optic products. Cable and accessories, such as connectors, splices, test equipment and transmitter and receiver equipment. YSA would sell their products when copper cable was not the best solution. The manufacturers only sold individual components, not systems. Scott recognized this as a massive opportunity. It took about six months to figure out that they needed to sell systems – and solutions – not individual pieces. It was obvious YSA had a chance to become not just leaders in one region, but across the USA.

Everything was moving fast and Scott believed he understood where the market was heading. YSA moved into fiber optics systems with maximum effort. They bet everything on it while still selling plenum rated cables.

It became obvious that few people understood how this stuff worked or why someone would want to use it. The fiber-optic manufacturers were scientists who started their own companies. They had scant concept of how to sell their products. They understood lasers, fiber optic cable design and how to inject light into fibers. They had little idea how to motivate a buyer, and teach them the value proposition of their solution. They must educate potential customers. They must explain why someone would install a fiber optic system. They must show what it would do for them. It also was obvious that it was no good if it was not a solution to a real-world problem.

Big Enterprise users, or leading players like NASA, all wanted a solution, not technology. He knew that if he told them all about how the fiber carried light or lasers worked or how to make splices, they would tune out. They would ask what does the fiber optics do to make my business better or my life easier? YSA needed to explain the answer based on their specific environment and needs.

Scott learned that the solution was relevant to the environment. Big petrochemical plants had long-distances to span. They also had to worry about lightning strikes. Lightning is bad news on long cable runs, frying the electronics. Fiber optics are immune to lightning. Distance limits of Ethernet were problematic. It can only travel several hundred feet over copper, insufficient in large plants.

As when he began, Scott found that customers were eager to help figure things out. Together they could analyze their specific needs and goals. He learned from them. This method of developing a customer base, of selling new products, and of building businesses would serve him throughout his career.

Token Ring Networking

At that time, IBM had created a copper cabling system just for their LANs. The LAN technology for IBM was called "Token Ring." IBM had the largest installed base of computer systems (both mainframe and mini-computers, such as IBM System 34, 36 and 38 which later was replaced by the AS400).

IBM invented Token Ring LAN technology to connect terminals more easily. To make the Token Ring work over the copper cable they invented the "IBM cabling system."

The IBM cabling system used a very expensive cable design for drops to each workstation or terminal. It had a *balun*, an electrical device that converts a balanced signal – two signals working against each other where the ground is irrelevant – and an unbalanced signal – a single signal working against ground or pseudo-ground – that allowed this fast 16 Mbps signal to travel over distances longer than 30 or 40 feet and still work properly. Token Ring used a protocol that was highly deterministic because they pass a token around the loop in a predictable time. Token Ring[21] ran at either 4 Mbps (IBM Proprietary) or 16 Mbps (IEEE 802.5). There was a 100 Mbps variant that never caught on. It was the ultimate LAN protocol and system, and very expensive.

Proponents of Token Ring advanced the deterministic vs. non-deterministic argument, but it was losing ground against TCP/IP over Ethernet. To be fair, Token Ring is technically superior to Ethernet in many regards save two; cost and ease of installation. It ran at 16 Mbps vs 10 Mbps, and the token-passing methodology meant that performance degraded much less under heavy load. In the real world, it performed much better than Ethernet when traffic became heavy. However, it was expensive, proprietary and poorly supported. The bulky cables and baluns meant it was difficult to install, and a problem in one termination could take the entire network down.

Most people using Token Ring went with the 16 Mbps version especially if they had many workstations connected. This was a high data-rate at that time. An IBM Terminal ran at 9.6 kilobits per second (kbps) or 19.2 kbps and a cluster of terminals connected via a controller on a "high-speed" 56 kbps circuit. Sixteen Mbps shared across multiple terminals was a huge jump in data rate and required new cable designs for everything. The biggest orders YSA had at that time were for IBM cabling systems, the Teflon-coated plenum rated version was like a gold mine. YSA came to appreciate LANs, which were popping up everywhere.

---

[21] https://en.wikipedia.org/wiki/Token_ring

Determinism ...

The question of deterministic vs. non-deterministic routing was a big deal in the early days of networking. Endless amounts of time and worry have been expended on fears of non-deterministic behavior. In hindsight, this whole issue would seem to have been an example of FUD (**F**ear, **U**ncertainty and **D**oubt), sown by IBM and the telephone companies to justify their solutions vs. much less expensive LAN-based services.

Determinism is often confused with *Quality of Service* (QoS),[22] especially as relates to congestion management. The two are very different. In computer science, a deterministic algorithm,[23] given the same input, always produces the same output. Saying a communications service is deterministic means that the transit time, or delay, required for every element of data from source to destination is predictable, or consistent. Two factors contribute to the delay. First, the propagation speed is about 130 miles per millisecond, not a serious factor unless you're calling someone on the moon. The speed of propagation remains constant and predictable.

Delays through any electronics are not so simple. In the AT&T digital TDM telephony hierarchy, the voice is sliced into little pieces, eight thousand slices per second. Those slices are represented by 8 bit numbers, (between 0 and 255) one number per slice. Those numbers are marched onto the wire, in lockstep with a rigid system clock, much like columns of soldiers marching in lockstep in a parade to the sound of a drummer. We call this technique *Pulse Code Modulation* (PCM) and properly done it sounds excellent. PCM introduces a few milliseconds of delay and again it remains constant.

A network where the delay remains constant is said to be deterministic.

Less rigid environments such as VoIP lack this consistency. We stuff numbers into an IP packet and send it over the Internet, through a few routers. There are many elements that can introduce delays, and those delays can be variable. One number may arrive in 10 milliseconds, a later number may be delayed 50 milliseconds, and other numbers may arrive in 100, 200 or even 500 milliseconds or may not arrive at all. This variability also sometimes called *jitter*.[24] There is no consistency or predictability. This lack of predictability means it is non-deterministic.

So, how do we cope? Ethernet of today is much faster, and full-duplex, thus network-slowing collisions are eliminated. Long-haul trunks are also much faster. We buffer the packets to smooth out the flow and become jitter tolerant. Delays still occur due to congestion, VoIP and Streaming sometimes sound horrible, with delays, digital noise, repetitions and other artifacts, yet, in the end, it works well enough.

It turns out that the whole "determinism bogeyman" was never the issue it was purported to be. QoS on the other hand is extremely important in carrying streaming media, and the service providers eventually abandoned the false concern of "determinism" and focused on providing real and effective QoS.

---

[22] https://en.wikipedia.org/wiki/Quality_of_service
[23] https://en.wikipedia.org/wiki/Deterministic_algorithm
[24] https://en.wikipedia.org/wiki/Jitter

The Rise of Ethernet

DEC's minicomputers started switching from using direct-connected terminals to using Ethernet as the way to connect terminals and workstations. Ethernet was disparaged as non-deterministic because it used a protocol that allowed collisions of frames and then retransmitted those lost frames when collisions occurred, thus introducing delays. That was called *Carrier Sense Multiple Access with Collision Detection* (CSMA/CD).[25] The data rate of Ethernet was 10 Mbps. The data rate is often called the bandwidth, though that's not quite accurate. The original Ethernet was half-duplex, meaning only one station could talk at once. If two "spoke" at the same time, there would be a *collision*, ruining the data, after which both would stop and wait a random time before trying again. The effective throughput was often much less when there were multiple terminals or nodes on one Ethernet segment, degrading rapidly as more terminals were added. Later versions of Ethernet would remove these restrictions, become faster and full-duplex, eliminating collisions.

Throughput is the amount of data transmitted versus the theoretical amount or the data rate. The total potential of signal carrying capacity is often referred to as bandwidth, which is the pipe size versus the actual throughput. The throughput is always less than the bandwidth. This concept of bandwidth and throughput would be a recurring concept Scott would deal with throughout his career. He first learned these concepts from selling cable to customers using computers and LAN systems.

Philip Sumners was a certified electrical engineer, and the "S" in YSA. Engineers are a special breed; they think differently. They use math and calculus to calculate how things work. They don't like gray areas, uncertainty, or things that cannot be calculated to perform a specific way. This is what makes them valuable. This is also what makes them difficult at times.

Running your own business and representing manufacturers is not like that. Everything seems gray, there's very little black and white. To sell products YSA had to find end-user customers, product distributors, and then convince the factory to give them good pricing and good deliveries in support of their orders. YSA had to do this multiple times each month to pay the bills and, hopefully, salaries.

If YSA focused on becoming representatives for fiber optic products that were used inside a building, or a campus, they could tap into the same market as the copper cable business. During the customer research, they would spot the opportunities and could recommend fiber optics when it made sense to the customers.

Customers such as NASA, large Enterprise companies on a campus, large petrochemical companies, and refineries needed to run cables in the hundreds of feet to thousands of feet between locations. They used it for low-voltage signaling, computer, telephone and energy management systems. (Low voltage signals, usually RS232,[26] are between +3 to +15 volts and – 5 to -15 volts.)

---

[25] https://en.wikipedia.org/wiki/Carrier_sense_multiple_access_with_collision_detection
[26] https://en.wikipedia.org/wiki/RS-232 - Voltage_levels

A system that used low voltage copper cable was often distance limited to about 300 feet, perhaps more if high-quality low-capacitance cable is used depending on variables. They were low speed and distance-limited. Fiber optics could change that dramatically by allowing data rates of many Megabits Per Second (Mbps) and reach thousands of feet, or even many miles, at those fantastic data rates without inducing errors. These systems needed to be able to transmit over cable without having more than 10 to the -9-power bit error rate. At these higher LAN speeds and longer distances, fiber optics proved by far the better choice.

The bit-rate decreased and error rate increased with copper wires over increasing distances. Fiber optic systems solved those problems and supported a high data rate and few errors over long distances. Fiber optics worked well for long runs in low voltage system situations involving computer and telephone signals.

YSA set out to be the best in the newly evolving field of connecting systems with fiber optics. They selected their own set of products to represent, developed their own set of slides and charged companies to attend seminars to teach them how to design and build a fiber optic system. They were successful not only in getting systems integrated into IBM and DEC communication systems located in high-rise buildings and campus environments but also to video and telephone systems, as well as programmable controller systems in refineries and big industrial plants.

Fiber optics was being used in LANs built by DEC and IBM. There was another implementation of LANs occurring in parallel with the DEC and IBM created LANs. A company called Synoptics came out with a way to make the LAN called Ethernet work over *Unshielded Twisted Pair* (UTP)[27] cable and shrank the bus cable down inside a box called a concentrator hub. This changed the game. It was still a half-duplex, shared media, but about 1990 the "hub" would be replaced with a "switch" and the IEEE 802.3i standard would be published and the interface would become full-duplex capable. With full-duplex switching, delay-inducing collisions are eliminated and the media supports simultaneous communications in both directions.

When Novell,[28] a company founded in 1979 in Orem, Utah, introduced their Novell Netware networking platform in January 1983, things started to change. Any small budget within a department level of a big company or a small company could afford and use Novell LAN software on PCs combined with Synoptics hubs and low-cost cable to connect PCs to each other via a Novell Server. PCs were appearing with *Network Interface Cards* (NIC), which let them connect to the twisted pair cable and back to the Synoptics concentrator/hub in a low-cost way.

The predecessor to Novell was a company called ARCNET, which dominated the LAN market in the early days of the 1980s. The new Novell LAN/Synoptics twisted pair cable solution dramatically lowered costs and changed the landscape in a few short years. A few years later Al Fenn would remark that the common RJ45 Ethernet connector was the RS232[29] port of the 1990s.

---

[27] https://en.wikipedia.org/wiki/Twisted_pair
[28] https://en.wikipedia.org/wiki/Novell
[29] https://en.wikipedia.org/wiki/RS-232

LANs were being installed everywhere at the department and small business level and were often making an end-run around host mainframes inside businesses.

It was a grassroots movement. And, it needed copper cable and fiber optics to make these systems work in offices that might be spread through a building or across a campus.

YSA didn't focus on fiber optics for long-distance telephone systems. In that market, big companies were buying single mode fiber and the suppliers for big well-established companies like AT&T were companies like *Corning Glass Works*.[30]

Corning invented and commercialized fiber optic fibers. Their fibers were used in all long-distance cables. They also were used in data-oriented fiber optic fibers. Companies like Siecor (an offspring company of Siemens and Corning Glass Works) sold most of the single-mode fiber optic cable from the late 70s until the early 90s.

YSA worked with new and evolving manufacturing companies. They were the first representatives for *Optical Cable Corp* (OCC),[31] a new startup. They were focused on making multi-mode fiber intended for shorter distances and lower data rates than single-mode fibers. The data rate for multi-mode cable was greater than 1 gigabit per second, which was extremely fast compared to Ethernet's 10 megabits per second and Token Ring's 16 megabits per second. Multi-mode fiber also allowed less expensive components like *Light Emitting Diodes* (LED) instead of lasers to drive electronics that carried the signals.

Scott met with Bob Thompson, the founder of OCC, at the fiber optic and LAN tradeshow called FOCLAN, in Las Vegas. He signed up to become Bob's first independent manufacturer's representative in 1984. Scott had pioneered plenum-rated data cables and had earned a good reputation for pioneering new markets that required technical knowledge with consultative selling techniques. YSA made the decision to focus heavily on the future of fiber optics systems and eventually developed a name in fiber optic systems for themselves.

YSA provided cable for LAN and campus environments to support all kinds of computer, telephone and data systems. It was a not a huge market in 1984. They were attempting to grow the market so more companies would buy and YSA could sell more products. There was no assurance that connecting LANS and telephone systems with copper and fiber optics would become a viable way to make money. Scott simply believed it was his best chance of making a good living at that time.

Many industry people thought it was very risky and too small a market to build a business around, much less make a living. In 1984 they were right. YSA struggled financially as they spent time evangelizing fiber optics in campus and high-rise building markets.

Scott hoped as the market grew YSA would emerge as leaders in the field and get a lot of orders. They were aiming to be the best at designing and installing fiber optic systems. And they were working hard to make that happen. He was certain they were on the right track. There was no certainty, however, that YSA would survive long enough for the market to blossom. He pushed on, hoping all the while to make something of it before they ran out of resources.

---

[30] https://en.wikipedia.org/wiki/Corning_Inc. - Corning_Glass_Works
[31] https://en.wikipedia.org/wiki/Optical_Cable_Corporation

HOUSTON, WE HAVE ...

(1984-1988)

A point came at which he knew they were going to build fiber optic networks. It's so easy to say that now. But at the time, people had no clue what a fiber optic network was. Such networks didn't exist. And when they did move – with great deliberation and amid warnings that what they were doing sounded crazy – they were largely going it alone. He'd been dealt a great entrepreneurial break-through. They were innovating at the dawn of the Internet Age, which was called the Information Age at that time.

Understanding the market and the opportunity in 1984 requires understanding the history of AT&T and the 1984 breakup of a seventy-year-old monopoly. The impact this had on the development of the Internet cannot be ignored.

The dissatisfaction with the telephone industry resulted in a series of lawsuits beginning in 1967 with the Carterfone case.[32][33] This was a manifestation of long-running dissatisfaction with AT&T's heavy-handed tactics which persisted for decades.

AT&T had always pursued a monopoly on telecommunications. By 1913, their tactics became unbearable and the government intervened. AT&T had been granted a U.S. Government sanctioned "Protected Monopoly" in the *Kingsbury Commitment of 1913*.[34] After some 70 years, this ruling was overturned and the market was deregulated. On January 1, 1984, the Kingsbury Commitment was voided. The AT&T monopoly was broken into *Regional Bell Operating Companies* (RBOC) known as "Baby Bells," with the parent company, AT&T, continuing in the new scheme as a long-distance carrier only.

The RBOCs were incensed. They saw AT&T as stealing the revenue-rich long-distance service and leaving the RBOCs stuck with marginal local service. An intense hatred developed over this issue that lasted most of a decade. This changed when competitive long-distance carriers, such as MCI and Sprint, entered the market. They were the first, but they were soon joined by many more. The revenue from long-distance service plummeted to effectively zero, bankrupting AT&T. In 2005 SBC Corporation purchased the remains of AT&T, incorporated it into their operation and renamed themselves AT&T to capitalize on the well-known name. Although the name lives on, the AT&T of old no longer exists. In 1993, Pacific Multimedia produced a compelling and entertaining documentary called "Hells Bells: A Radio History of the Telephone"[35] about the history of the Bell Breakup. Recommended.

---

[32] https://web.archive.org/web/20150120021035/http://www.uiowa.edu:80/~cyberlaw/FCCOps/1968/13F2-420.html

[33] https://en.wikipedia.org/wiki/Carterfone

[34] https://en.wikipedia.org/wiki/Kingsbury_Commitment

[35] http://town.hall.org/radio/HellsBells/

In 1984 there was no competition for local service. There was only limited competition for a slice of the long-distance markets, led by MCI[36] and Sprint.[37] Both were well-established telecommunications companies prepared to jump into the deregulated space, and many upstarts followed on their heels.

Local metro networks were new, and not considered profitable. Local service had always been a money-loser, subsidized by the long-distance revenue. Turning this perception around and making local service profitable was not believed practical. Separated from long-distance subsidies under the new regulatory scheme, profitability was considered unlikely.

Scott was selling plenum and PVC telephone cable when he became familiar with the then new idea of *"long-distance resellers"* who emerged in the deregulated market. He looked at getting into the long-distance reseller market in Lufkin around 1983-1984. It seemed very risky and required a large investment. It needed a minimum of $500k to get started. He passed.

But the research process was educational. Scott learned things that would soon become the core of a new business. He learned how long-distance switches connected to other long-distance networks of switches and the T1 circuits used to connect between cities. This system is what allowed resellers to remove traffic from the traditional switched network. The resellers could offer long-distance service at lower cost, taking advantage of inefficiencies in the established phone companies. Monopolies never build efficient organizations, and there were tremendous inefficiencies indeed in these hold-overs from the days of monopoly.

This was also why resellers provided local phone numbers to allow people to dial-up directly to the long-distance network in each city as a local phone call, which reduced the cost of originating a call from that city. If you wanted AT&T, for example, you dialed 288 as a local number and receive a dial tone then dial your long-distance number, effectively bypassing the local carrier. Each long-distance provider had its own "Access Code" and callers could freely choose which one they wanted.

These concepts became very important in understanding the way the long-distance switching network functioned in the new deregulated environment.

Scott didn't know at the time how important this knowledge would become. When building a fiber optic local loop company, a key business niche would become dedicated access from a building to the long-distance carrier of choice. Understanding the purpose, function, and capacities of the existing switching networks became critical to the new business. It would be the business focus. The time spent researching long-distance reselling paid off in an understanding of the key concepts that would give structure to a metro fiber network.

Since YSA had decided to focus on fiber optics for LANs it became clear that they had to sell all the components of the entire fiber optic system to make money. Scott decided that seminars were the way to educate potential customers and win them over. He wanted them to view YSA as the leader in fiber optics so they would purchase their products and possibly choose YSA to install the systems.

---

[36] https://en.wikipedia.org/wiki/MCI_Communications
[37] https://en.wikipedia.org/wiki/Sprint_Corporation

In 1984-85 this was all new terrain and a certain measure of fear and uncertainty accompanied each step forward. They were very small and lived from commission check to commission check. They had no capital to invest and no room for failure. They were cautious of taking on any project that didn't make money. They were surviving months when they were so strapped for cash they could barely pay the bills. These tight financial times in which sales revenues barely covered costs endured several years.

Philip and Scott became evangelists for building the complete fiber optic solution for systems applications in a building on a campus. They developed slideshows and gave day-long seminars that covered fiber optic cables and systems from A to Z. They covered everything a business needed to know about connectors, splicers, test equipment, modems, transceivers, and multiplexers. Then they attempted to tie these components together to demonstrate how they functioned as end-to-end solutions. The customers found it hard to believe light carried information down an internal mirror fiber smaller than a hair made of glass.

The slides[38] YSA created to teach end-users about fiber optics are something to behold. They didn't have PowerPoint or other software to make their slides. They paid graphic artists to take their ideas and turn them into 35mm slides which were presented using a projector and screen. The slides were indispensable aids in helping people understand cables, connectors, splices and how electronics convert analog and digital signals from electrons to light and back again after traveling over fiber optic cables.

It turned out these also were essentially the same components needed to design and build a metropolitan fiber optic network. It soon dawned on Scott that he could use and capitalize on his experience connecting LANs, video, data and telephone systems over fiber optics. It was, after all, what he'd been doing already within campus environments. The Rohm and Haas Chemical Company IBM system had cable runs over one mile in length. That was as far as one end of downtown Houston to the other end of downtown, a point which would become important later.

When beginning the wire and cable business with plenum-rated cables, it was an adventure doing something novel in which he had no experience. And, as always, the valuable experience at CECO lay behind that success. The CECO experience instilled the confidence and willingness to trust his instincts when it came to trying new ways of doing things. And he would again call on that training and confidence to pursue his next high-risk venture.

---

[38] http://www.fscottyeager.com/slides

THE FIRST METROPOLITAN FIBER OPTIC NETWORK

The transition from selling fiber-optic cable to installing a metropolitan fiber-optic network was a huge next step. The formation of the new company and development of the business plan was a bold, risky step. NCI became an important player in the drama unfolding in the field of data communications, as described in Scott's original first-person account.

[---]

By 1985 YSA had sold and installed a great deal of fiber optics systems for voice, video and data on business campuses. Several of my YSA customers asked us to connect buildings across the street from each other. This involved installing cables that passed over the rights-of-way of the City of Houston. This prompted me to consider whether an opportunity existed to create a new kind of company – one that sold a service over the fiber optics network versus selling equipment to our customer end users.

Philip Sumners and I had by now become the leaders in fiber optics I'd hoped we'd be, at least in our geographic region. We'd become especially skilled in educating people in how fiber optics could work and the value to businesses that used the technology, especially large companies with large campus environments. In a campus environment, the computer data center was in one building and the users were spread out all over. And they tended to have different requirements. Some, for instance, required connections to several types of computer systems — an IBM 3090 here, or a DEC minicomputer there — to carry out different aspects of their business.

Using our deep experience associated with connecting voice, video and data between large buildings and large campuses as a backdrop, we were helping our clients see how it made sense to deploy fiber optics to interconnect all end users between multiple types of computing environments. We also demonstrated the limitations and costliness of copper, how more effective stand-alone LANs like Novell's were compared to Ethernet alone, and how extending Ethernet over 500-feet became problematic. On the other hand, Fiber optic systems could allow cable runs to extend up to a mile without signal distortion or loss.

Our new business model was vastly different than a manufacturing business like CECO's. At CECO we worked with a unit cost versus a unit profit. Manufacturing required that we sell more products each month to a new or existing customer – or we realized zero revenues. The idea of monthly recurring revenue is foreign to manufacturing. Our new model repositioned fiber optic systems within delivered services added. The customer paid monthly fees to use the capabilities created by the fiber optic technology layered on top of the physical fibers. Among other benefits, this meant we'd no longer be stuck on the merry-go-round of selling more and more new products each month to more customers.

The phone company's business model was to sell dial tone and long-distance services. It was a monthly recurring revenue "utility" type of business. And it held the potential to generate huge profits, but it was a model typically available only to monopoly utilities. That is until the monopoly was broken and competition from other long-distance companies started to break the monopolistic stranglehold on monthly recurring revenues. Expensive long-distance calls became an inexpensive commodity under the pressure of competition.

To get in the game and make a profit required risking huge capital costs up front, then there would be fixed but minimal operating costs for employees and rent, with the amortization of the infrastructure over time. It was more sophisticated in some ways than what I was dealing with, but also not unlike bidding on a systems integration project from a cost perspective, which I was familiar with. The way you made a profit was not to get all your costs back in one big invoice but to reel it in over time. You expensed things, amortized and depreciated. If you got the right revenues using the infrastructure you stood to make a lot of profit.

This was when I figured out how much money a monthly recurring revenue business could make once a certain number of companies or customers were paying you so much money per month. The Monthly Recurring Revenues (MRR) model, or the Rule of 78, that was once the domain of the power company and Bell monopoly, became the norm for new competitive phone companies. Once you reached the cash flow break-even point and could pay the bills you were well on your way to a cash cow business that virtually printed money – especially if you did not have to pay for more equipment (except on a minor incremental basis) as you got more orders and more customers.

This was a huge revelation. And I found I could model the whole thing using Lotus 123 spreadsheet software on a cheap PC. Soon I would be sharing what I was doing with others, including investors and the City of Houston.

[---]

The MRR model only became practical for businesses in the mid-80s when financial analysis moved from "big iron" mainframes and big information technology departments. Thanks to PCs, control moved into the hands of novices who were suddenly in possession of the necessary computing and analytical software.

Scott could use a PC and Lotus 123 software to model the financial structure of a whole new way of offering voice, video, and high-speed data services over new technology and fiber optics. He was, in effect, creating a new type of infrastructure to deliver such services – and it reached all the way to the buildings in which Enterprise customers were located.

He knew the costs of such a business inside and out. He knew the costs of the multiplexers, the fiber optic cables, the splices, the patch panels, and the connectors. He knew the cost to dig the streets up, to install conduit, and to pull cable into the conduit. By late 1985 He was accurately estimating the cost of building a network to buildings in downtown Houston.

To estimate revenues, the first step was to determine how much they could charge for the services. He went to Austin and looked up the SWBT tariff filings for DS0, DS1 and DS3 circuits. These were the building blocks for the telephony services of every local telecom monopoly.

Prices for within the *Local Access and Transport Area* (LATA) — called Intra-LATA services — were regulated by the *Public Utilities Commission* (PUC). A similar tariff with the *Federal Communications Commission* (FCC) regulated Inter-LATA connections (between the *Interexchange Carriers* (IXC) and the LATA), and between local monopolies, in other cities or states.

IXCs included AT&T, MCI, and Sprint. Of course, AT&T alone generated more than 80 percent of all long-distance revenues in the mid-1980s. It wasn't quite a monopoly, but it was certainly the dominant player, and often acted as if the old monopoly were still in force. The RBOCs like SWBT had to pay to interconnect to the IXCs.

Of course, YSA had to pay those prices, too, and modeled those monthly recurring rates to estimate how many circuits of each type must be sold for the business to make money. In the mid-1980s, the ability to use a Lotus 123 spreadsheet in this way was novel, and a tremendous boon to a small businessman trying to model financial viability. Demonstrating technological viability was one thing. The plan had to make economic business sense as well.

Before this, constructing such a model without paying immense sums to a consulting or accounting firm would have been impossible. This was the power of the PC revolution. One person alone with inexpensive tools could perform complex modeling. One person, with an inexpensive computer and publicly available data, could model a hypothetical new business.

Once the model demonstrated it was possible to make money, then the launch concern became, whether the ramp up curve of revenues versus costs was too high? Could revenues grow faster than costs so as not to rack up huge losses over a long period?

Not only was he modeling a new company, but one in an industry that was dominated by an incumbent, government sanctioned monopoly. The prospect of creating a new entrant in this hostile, monopoly space was daunting.

The model indicated that it was a viable business plan. Based on the favorable model, Scott decided to pursue this huge opportunity even though as a small fiber-optic system integrator YSA lacked experience as a services business. Of course, no one else had done it yet either.

Opportunity beckoned.

Under YSA, Scott had sold a FOTEC variable attenuator to Rockwell Collins, Inc.[39] which made a *Time Division Multiplexer* (TDM) that delivered DS1s, T1s or DS3s over a 565 Mbps (12 DS3s, the maximum on one fiber at the time) fiber backbone and went 50 miles without needing a repeater. Fifty miles meant the signal at the far end would be less than one eight-thousandth of the starting signal, the rest dissipated in the glass. It needed a powerful laser to reach so far.

---

[39] https://en.wikipedia.org/wiki/Rockwell_Collins

When SWBT started putting fiber in place between the *Central Offices* (CO), they encountered problems. The CO were only 5-to-10 miles apart. At that distance, the light hitting the receivers would be over 500 times "brighter" than at fifty miles. The lasers were much "too hot," that is, put out too much light and overloaded the remote receivers. Fotec, Inc.[40] (Now a part of Fluke Networks, Inc.), had the only variable attenuator on the market. The concept of introducing losses in a signal path on purpose was a new and almost scandalous idea. The goal was to go long distances without needing a repeater, not short distances.

It made sense to build fiber optics between CO[41] because there are a very high number or circuits between those locations. Fiber optics carried 672 DSO times 12 DS3s, or over 8000 voice circuits, on 4 fibers up to 50 miles with no repeater and no loss or signal quality. It was very powerful for interconnecting COs.

This made it clear that fiber optic systems could be and in fact were being deployed in the metropolitan market. YSA was dealing with one of the key manufacturers, selling them the FOTEC attenuator. It seemed reasonable that one could install fiber optic systems inside a city. YSA was watching it occur. Scott knew the engineers at Rockwell Collins were good and their product was good or they could not sell to SWBT.

SWBT had no incentive to build a fiber network within a city to meet competition. There wasn't any. They had no need to compete against themselves. They were, however, installing it to make their network more efficient and operate better between their central office switch locations. The growing need for telephone circuits between COs was outstripping the capacity of copper.

Scott was a telephone man in college, visited the CO and observed the switches and operations, thus felt a sense of how their network needs between COs were different than connecting to the end user buildings. They were still a monopoly. No one imagined competing with the local monopoly in 1985.

To explore whether the financial side of the equation was as viable as early research indicated, Scott turned to a friend, Larry Koonce. He explained the idea and Larry was willing to look at the concept. He had been with Arthur Andersen[42] and he'd also served as CFO for a small oil company. Best of all, he had experience in evaluating deals.

Larry thought he could "get the money" to build a multi-million-dollar network if the idea proved feasible. Building a metropolitan fiber optic network was a radical idea, and competing with "TPC" seemed an insane one. Feasibility was not assured. Larry estimated they'd need $3 million in 1985 dollars (about $6.75 million today) to build the minimum network. Realistically, the new company would probably need up to $10 million ($22.5 million today) to build it and climb that long ramp to profitability. This seemed an impossible amount of money. Scott feared debt and did not owe any money, it was scary to contemplate raising that kind of money. It seemed an insurmountable hurdle without Larry Koonce saying it was possible.

---

[40] http://www.lightwaveonline.com/articles/2001/03/fluke-networks-acquires-fotec-inc-53467942.html
[41] https://en.wikipedia.org/wiki/Telephone_exchange
[42] https://en.wikipedia.org/wiki/Arthur_Andersen

Call Me Crazy

It was a crazy amount of money to raise. No question about it. And the whole venture capital tech boom had not bloomed. What's more, there was no precedent for a new company attacking a multi-billion-dollar monopoly – and doing so on *their* ground.

Scott describes his thinking:

[---]

I had grown up watching a show called *The Millionaire* in which everyday people were surprised with a $1 million gift. It was a widely-watched program because the average person considered that amount of money unattainable. I certainly knew the feeling. I had no notion of how one went about raising that kind of money. It was a lot. Also, who would be crazy enough to invest in a company that was going to try to compete with a multi-billion-dollar monopoly like SWBT?

In 1986 the idea seemed crazy to most of the people with whom I shared it. It seemed even more far-fetched if they knew I had no money or experience at raising money. Oh, and there was the matter of no one ever inventing a service to compete with the local phone monopoly. How could I do that? Was it legal? Even doable? And if it was, well then, why had no big company done it already? Everyone seemed to have questions.

"I thought of it," I'd answer. "And I can see no one is doing it. Yet there's a clear opportunity and it meets a need in the market. So, I figured it out. And now I'm about to do it. "

I remembered how crazy it seemed to start YSA, our rep firm, with almost no revenues when I first started out. And I remember how crazy and uncertain it was when I decided to become a manufacturer's representative for fiber optics. No one knew how to use it, or even for what purpose it was best used, and YSA had no experience yet with fiber optics.

This new venture was just the "next crazy thing" to do — if we could determine it made technical and financial sense. And all my research was telling me it made sense.

Like I said, I was only familiar with financing the growth of a business through increasing revenues, not using other people's money, or OPM. And OPM concerned me. Wasn't it taking on a huge burden – not to lose that money? It was different if it was your own money, I thought. But Larry Koonce said it was common for people to raise money from investors to build new businesses, though they often then gave up control and had to stand in line behind the investors when it came to getting profits out of the deal.

I had determined that we could physically build such a fiber optics network. My engineering buddies at Rockwell Collins had confirmed it was not only technically feasible, they said I could buy their stuff and do it like SWBT was doing it between their CO. I felt confident we could win over customers and make them happy. That made me comfortable enough to proceed.

I used PUC and FCC tariff files by SWBT as a guide to help me price the number of DS0, DS1 and DS3 circuits we would sell each month. We wanted to see how long it would take to get to a break-even cash flow. It was assumed that if we did get customers to pay for a monthly recurring service that our operating costs would be low enough that we could make money eventually. This would not happen until the fixed costs were amortized and paid for while the variable costs were added in. The variable costs, like the unit cost of the circuits and the labor to install and maintain them, were then taken into consideration.

I credit the Lotus 123 spreadsheet software that we ran on the IBM with enabling me, as a small business, to think big by allowing me to figure costs and go for making a profit. It was a godsend to a person like myself who had a degree in biology and had never taken a business course in his life. I had helped build a business at CECO. I had built a wire, cable, fiber optic sales representative business. But this? This was different. This was big business and took another level of expertise. I needed a financial person like Larry Koonce to look at the deal, knock holes in it, tease out any fatal flaws.

Larry Koonce and I went to a nationally franchised bar on Bissonnet Street close to my office and I drew a network on a napkin to explain the way companies I was familiar with wanted to interconnect sites. I told him that Texaco and some other customers had said they would buy the service from us once it was real. All we had to do was build it. I told him why I thought it would make money.

Larry Koonce took my ideas and looked for the flaw. If he couldn't find one and we moved forward we would decide what the deal would be between us later. I wish in hindsight I'd saved that napkin.

At first, I was a little embarrassed to bring up such a wild idea. After all, what I was proposing would cost millions to build before we even brought on our first customer. That was way more than I could ever imagine raising on my own. I still could not believe there were people who would invest in things like this. But Larry reminded me people invested in drilling oil wells all the time. And, he said, it was understood there might be dry holes, that investors might lose. This gave me some hope we could get the money.

Larry Koonce reviewed the models I'd created and provided a real financial evaluation from his perspective, changing the models as needed. The financials showed that there was indeed a viable business that could make huge profits — if we achieved a critical mass of customers for a specific footprint of fiber optics networked to buildings in a downtown area. And then? We'd expand, of course.

By May 1986, Larry and I had decided there was a business and it could make money. We had the cost figures and the revenue numbers for circuits figured out. And we'd created a "straw man" business plan to see if the company would ever make money.

I used my contacts with Rockwell and NEC to get cost figures on fiber optic multiplexer equipment and I could get costs on cable, connectors, splicers, test equipment and labor to install all this. We could estimate the number of people it took to run it and manage it. So, we had the costs and the revenues and timeline projections. This was based on my fiber optic systems experience, but still, we had never operated a network and sold services over the infrastructure, so there was still some guessing going on.

Larry Koonce was convinced we had a business. However, many people who were friends, or business players in the industry, still thought it was "way too crazy."

We decided to go for it.

[---]

NCI - FIBER OPTIC TRAILBLAZER

(1985-1989)

With the technical and financial indicators reading positive Scott and Larry took the next step. They wrote a letter to the Texas PUC in Austin on May 5, 1986. It was important to discover what regulatory hurdles, if any, existed. They initially named the new company Digital Services. They liked the generic nature of the name. They soon discovered that name had already been claimed and came up with another -- *Network Communications Inc.* (NCI) Scott liked the new name even better. It clearly emphasized networks, which were to form the core of the new business. Although, from the start they also were thinking that in time they would move to the use of multiple technologies, not just fiber optics.

Scott explains:

[---]

I saw it as extending, expanding and innovating on the things we were already doing at YSA, beginning with connecting a company's buildings on a campus. But with a significant twist. We would reach beyond offering the physical platforms or infrastructure companies needed to interconnect voice and data between buildings. Now we were going to push fiber optic cable to connect buildings off campus, from across the street to across town. And, in an innovative step I knew of nobody else then taking, we would start selling interconnectivity as a service. Installing equipment to create the services versus selling it to the customer to install on their own fiber would be the big shift and become our new purpose. We would own and operate the fiber and the equipment and sell rights to our capacity on our network.

The notion for the new business arose organically like the best business opportunities do. My philosophy throughout my career has been that you need to be able to recognize them when they appear and then be equipped – thanks to experience, training, and solid critical thinking skills – to see what can be made of them. The new possibility, in this instance, cracked wide open when some of our YSA customers asked me if we could connect buildings which were not on their campuses.

Running anything via public rights of way was only allowed if you were a public utility such as the power company or the telephone monopoly. It was beyond a small fiber optics and cable company like ours to do that for clients one project at a time. The thought of fixing this problem for multiple companies in Houston in a uniform manner was appealing. I began to consider how we could cross public streets to other buildings to create fiber optic LAN extensions, IBM extensions, and even, here at the start of this new era of telephone deregulation, extensions for voice services.

The idea to build a service company was the natural result of responding creatively to customers asking me to provide a solution only a systems integrator could design.

There were many challenges, of course.

To begin with, we were not a utility company and controlled no right-of-way. Nor did we know how to get the right-of-way, or even if it was possible to lay fiber in the conduits that passed under Houston's public streets. No matter. Looking back, the key was that I had started thinking in late 1985 and into early 1986 that this could be a new business opportunity.

Before I knew it I was working on figuring out the kinds of services that could, or should, run over fiber optic networks. I could sense the outline of this service company emerging from our operations.

Fortunately, I was already familiar with the workings of the Texas PUC and its staff, because of my earlier experience in the telephone plenum cable market. I was soon heading their way for answers to some of the questions I'd developed for this new service provider business. I stepped into the regulatory thicket confident I could at least understand, and probably figure out and surmount, any problems that might stand in the way of realizing my vision for what I wanted to do next: Build a fiber optic network in the Houston metropolitan area.

I understood how to work with the commission staff. And I knew my way around the commission's library. This, I knew from experience, could prove helpful. When you went to the staff with well-researched questions they were more inclined to help you.

I'd first established friendly contacts within the PUC a few years earlier when I got caught up in changing the tariff on plenum-rated cable. (Telecommunications tariffs are basically contracts between the public and a telecommunications service provider that are typically regulated and approved by public utilities commissions. Tariffs include rates, fees and charges service providers levy to provide services and to recover their capital infrastructure costs.) The issue was simple. I'd discovered that SWBT was charging consumers a $2 per foot tariff for cable that I was selling for 25-cents per foot. Southwestern Bell, obviously, was overcharging. Once I proved the market price was much lower than the tariff, a PUC engineer helped change the tariff. I was pretty sure she did so because she quite rightly believed the PUC had been duped by Southwestern Bell.

The tariff matter was the sort of thing we were dealing with in 1984 when we were also selling a lot of plenum-rated telephone cable. We'd teamed up with Joe Boscov, the founder of cable manufacturer Berk-Tek Communications Cabling. Berk-Tek was coating copper cables used in plenum spaces with Kynar, a polyvinylidene fluoride polymer sold by the Pennwalt Corporation. It was not only an innovative product, it also allowed us to sell our version of plenum telephone cable far more cheaply than the Teflon versions sold by other companies. And we sold a lot – millions of feet in 1984-86. And we depended on those sales to survive while cobbling together our fiber optic system integration business.

Following up on a letter Larry Koonce and I wrote on the Digital Services Letter Head to the General Council at the PUC, I visited Don Price (who was a common carrier technical staff person) at the PUC. The response said we were a non-dominant common carrier and we thought it said we were not subject to regulation but did need to register. We knew each other from my plenum cable lobbying days. What Don said instantly strengthened my commitment to the course we were embarked on. Don said if I wanted to build a fiber network he would not regulate it. Don added, with a sort of verbal wink, that to regulate us would be tantamount to regulating a microwave link. And we both knew the Texas PUC did not regulate microwave links.

This was great news. We'd essentially just been given confirmation that we could legally build a metro network, at least per the rules of the commission. We already knew the FCC had no jurisdiction over municipal rights-of-way and did not regulate local businesses unless they crossed state lines or cut across a LATA boundary. Thus, we would not need approval from the FCC, either.

At this point, it all struck me as a little too easy and straightforward. It made me nervous, frankly. Why wasn't my fiber optic metro network idea already being developed by industry players elsewhere? It would remain a nagging question for months. Surely, someone somewhere in the country was pursuing something similar?

Our new business plan continued to take shape as we made further discoveries during this period. We were realizing, for example, the importance of explaining to customers the concept of point-to-point versus switched solutions. Circuits were point-to-point things. They went from "location 1" to "location 2." And they were on "all the time," until the service was turned off. This "always on" state is what distinguished DSL from a dial-up modem, by the way, though neither service existed at that time. This was a key concept in selling services to customers. They wanted to go from their building to their long-distance carrier, or from building A to building B, or to wherever they had employees, interconnecting telephone systems. Sometimes they wanted to interconnect computer systems using telephone circuits. The point-to-point solution was compatible with computer systems, certainly, although it also forced customers to convert protocols to use the telephone network to interconnect computers.

[---]

To accomplish point-to-point networking meant going beyond merely utilizing the existing infrastructure of existing T1 and DS3 circuits. T1 circuits were a Bell Labs-created carrier signaling scheme and the North American Standard in the telecommunications industry for the transmission of voice and data between devices. The first circuit level was 64 kbps, or some subset of it, such as 9.6 kbps, 19.2 kbps or 56 kbps. These were also called DS0's. A DS0 is the basic PCM voice channel in digital telephony, 8000 digital 8-bit samples per second.

A classic T1 circuit combines 24 Time Division Multiplexed DS0 channels, with each channel getting an isolated time slot. The data rate for a T1 is 1.544 megabits per second (Mbps). The T1 is also known as a DS1. The T1 is the actual line, the DS1 is the signal itself. A DS1 may be delivered over microwave, fiber or another mechanism, or it may be delivered on a copper T1 line. When DS1s are carried on DS3 circuits, 28 DS1s are multiplexed into a 44.736 Mbps stream. Thus, a DS3 carries 672 multiplexed voice channels. At the time, DS0, DS1 and DS3 circuits and this Time Division Multiplexing (TDM) technology fell wholly and solely within the province of the telephone company, and this would remain the case until the late 1990s when competitive carriers flourished and Internet traffic swelled.

A critical weakness of this system is that it is based on copper cabling, and copper limits the distance a signal can travel. A T1 is limited by various factors depending on the type of cable and type of physical interface. Typical limits quoted are 3000 feet over coaxial cable without a repeater, although up to 6000 feet are possible in some cases. When you combine 28 T1s you create a DS3, or T3 circuit, which provides 672 channels at 64 kbps per channel bandwidth (or 28 T1 channels of bandwidth). But a T3 can only travel 450 feet without a repeater over copper cable. Greater distances require repeaters or alternative (fiber-based) solutions.

The demand for more of these circuits was beginning to grow exponentially in the post-deregulation environment amid the competition to provide long-distance services and provide computing connections to Internet-reliant businesses. The existing copper T1/DS3 infrastructure could handle speeds of 1.5 to 45 Mbps.

As companies sought faster speeds, fiber became the cabling of choice, with speeds ranging from 5 to 100 Mbps. Unlike DS1/DS3, which enjoyed a near ubiquitous footprint thanks to telephony, the footprint for fiber was yet to be created in fiber optic metro networks.

The time was ripe for a company like NCI, with experience in fiber, to take advantage of the situation. Conditions were ideal for creating fiber optic networks in metropolitan areas to connect buildings and companies together or to the long-distance carrier of their choice.

Also, now the cost of long-distance minutes was by far the greatest expense for networks, and it, therefore, represented the greatest digital challenge to companies from the early 1980s through the late 1990s. It was advantageous that this is where most commercial interests were focused. While others chased riches in the Long-Distance market, NCI could blaze new ground in an unrecognized niche where there was little or no competition.

This emphasis on point-to-point telephone circuits would help ensure that the PUC would not get nervous about there being a local dial tone issue involved in NCI's proposal. NCI had no intention of becoming a local dial tone phone company. They wanted to show clearly that Network Communications Inc. was not just another competitor for local loop circuits competing with the phone company monopoly. NCI sought to stake out a claim as an innovator, creating new metro connections using LAN or IBM native protocols and native computer data rates over fiber.

INVENTING METRONETWORKS

NCI was headed in a new direction, developing and promoting the radical concept of connecting LANs between buildings. This all occurred over optical fiber, which, significantly, also continued to accommodate telephone circuits. No one else was talking about doing this.

Scott describes his thought process in his notes.

[---]

I say I "invented' this networked services idea, not out of some exalted sense of accomplishment but with the satisfaction of knowing it was born from personal experience. No one gave it to me. And I didn't borrow the idea. I was, in fact, oblivious that others might be moving in the same direction. I just figured out that I could build, own, and operate a metropolitan fiber optic network. And I assumed, correctly, that high-speed data services would be something companies would readily buy. The whole notion was based on my YSA experience connecting buildings dropped down here and there over the large expanse of an industrial campus. We'd connected buildings for a petrochemical company as well as Houston's Johnson Space Center, for instance.

I wasn't inventing a new technology, it's true. But I was applying technology in a new way. LANs already existed, but only inside buildings, or between buildings on a campus. Now, we were about to help LANs leave a building and run to other buildings in a city. This was no different to me than going several miles between buildings on a sprawling campus like Johnson Space Center's. It was all about connecting Ethernets between buildings. The space center just happened to own the property so NASA could dig up the streets and pull the cable. If the space center did not own the right-of-way it wouldn't be possible to interconnect those buildings.

Again, our plan was to provide data services that leveraged fiber optic networks. So, to differentiate Network Communications from other companies, we took care to explain and discuss how our special services differed from and were not offered by the phone company. I explained to Larry Koonce, the PUC, and others that we were especially different in our emphasis on "end user to end user" links between a single customer's buildings and a second service, which was connectivity from the building to their long-distance carrier. We would offer both voice and data services.

Networked computers were the future, and the protocols and speeds that computers required to interconnect were different from the phone company network.

I was intent on getting our data services company up and running as soon as possible, fearing others might also spot the opportunity and leap ahead of us. Boy, did I misjudge things. It would take years to get the industry to understand and it the new millennium would arrive before for the average person realized what a network connected computer was. Today, most people know what Ethernet is and how it's used. They know that connecting to the Internet leads to content and applications.

But initially, very few people understood what I was talking about. In the mid-1980s, it was one thing to create a new local phone company that sold DS0, DS1 or DS3 circuits to connect businesses to a long-distance company. The new companies that emerged with Telephony experience were focused on using fiber optics for voice communications, specifically using fiber for delivery of access to long-distance carrier networks. The industry players did not know about or care about the data applications customers were considering.

This was a good thing, of course, since we hoped to avoid being dragged into the dial tone quagmire. Our activities clearly positioned us in a new realm of data services, the first of their kind. And our regulatory agenda was to differentiate ourselves from the local telephone monopoly. We were going to be doing things that were different from what the local phone company did.

At the same time, we had to be prepared to show how Houston businesses would benefit from this technology and grow their business. Networked computer systems could help run their businesses more efficiently. My days at CECO, and the work we did at YSA to internetwork multiple types of businesses with fiber optics helped. I would describe for people the business processes that existed in companies I worked with and correlate efficiency gains to the willingness of these companies to network computers to end users.

Enterprise customers who had IBM or DEC computer systems were very interested in data connectivity, but especially those that had LAN requirements. The whole point was to interconnect employees and allow for shared resources and applications. The need to go faster and farther was intrinsic to the need for LANs. And, of course, the greater the number of users the more bandwidth needed. Phone company bottlenecks were holding back this optimum connectivity and efficiency.

Fiber optic metro networks could help solve the problem. And we were intent on showing how it could be done in Houston. After all, downtown Houston was only several miles across, an area no bigger than large industrial campuses YSA had already networked in Texas and Louisiana. In fact, some of the campuses where we'd installed equipment were physically larger than downtown Houston.

[---]

CONVINCING HOUSTON

Ultimately, NCI discovered that what was needed the most was a franchise that allowed digging up streets or using existing conduits owned by *Houston Lighting & Power* (HL&P) , the local utility company. When Scott spoke to people at the power company they said that unless NCI held a City of Houston franchise and functioned as a utility they would not permit NCI to share their conduits.

Larry Koonce found a man working on a land deal for an airport west of Houston who offered wise counsel on how to approach the city. He cautioned them to make an inquiry through a lobbyist they should task with determining if the city would outright oppose the franchise request. The advice of the law firm that made the inquiry was heartening: Go in the front door and ask for the franchise.

In the process of identifying the proper city government people to approach, NCI discovered there was a newly formed city franchise department. It was encouraging that this new department had been formed just as they were seeking a franchise for a new kind of utility service company. Scott hoped there would be less bias and a greater openness to new ideas.

Initially, city staff did not know what to make of NCI. It was unheard of in 1986 for an upstart fiber optic franchise to compete with a phone company. Franchise department head Susan Hodge asked "Why should we give you a franchise since we get all the revenue from the phone company anyway? Wouldn't we be robbing Peter to pay Paul if Houston granted a franchise to your company?"

People did not yet appreciate the possibilities appearing for companies other than the local telephone company to develop new services within what's called the "local loop" in the telephone communications infrastructure. The local loop, sometimes also called "the last mile" or the "subscriber's line," links the customer to a telephone company. There was now a growing need for new network configurations within that last mile. A network that could serve a rapidly changing world of business communications accelerated by the widespread use of computers.

Scott told Susan that fiber was a universal voice, video and data cabling medium. He emphasized the need for high-speed data competition in the marketplace. And that NCI would be using fiber to create new services not available through SWBT. The city was now entering the Information Age. Wouldn't it be a good idea to create competition in the local loop through innovations brought about by new technology?

Scott brought samples of fiber optic cable into these early meetings so people could touch and see the actual cable. He wanted to get them past an abstract mental image, to connect physically with the cable, to be able to see how multiple fibers could allow a cable to simultaneously carry robust, pluralistic streams of voice and data.

Scott believed the data focused LAN connectivity aspect of the plan was the key to success. Others, including Larry Koonce, thought he was nuts. Maybe he was, for that place and time, but if so, then it is typical of innovative thinkers anywhere, or anytime. Whenever one challenges the status quo, the status quo challenges back.

They thought he should stick to telephony services, which were much better understood and more in demand. Scott felt the combination of services was much more powerful and essential to get the franchise. A head-to-head competitor to the monopoly phone company had little chance. Scott also believed strongly that these innovative kinds of services would bring the network enough traffic to make it profitable.

A city staffer named George White was researching the NCI proposal. He was a very honest and straightforward person. He was also auditing the phone company and the power company to see that they paid the franchise fees they owed. Scott offered to help, based on his experience with the PUC. Perhaps, as with the Texas PUC, he could shed light on certain tariffs. Was the phone company collecting revenues but not paying the full city fees due? The experience looking at tariffs for plenum cable pointed him back to the PUC and FCC libraries where he again looked up specific SWBT tariffs.

Scott focused on switched access revenues. This was a new revenue source for the phone company that appeared in 1984 with the breakup of AT&T. The local phone company received a switched access fee of "so much per minute per long-distance call."

Questions & Answers

George revealed the fees being paid by SWBT to the city based on its total revenues. Then Scott went to the PUC and researched how much revenue SWBT was taking in in total across all areas and all categories. That figure covered five states. He had to break out the population and customer base for Texas to estimate how much of that revenue SWBT was deriving from Houston alone. The upshot: He found that SWBT was not paying Houston its proper fee. Scott illustrated for George that Southwestern Bell was eliminating local switched access revenue from the formula used to calculate the fees paid the city. It turned out the city was missing millions of dollars per year.

This, of course, also provided another good answer to Susan's big question. "Why should we give you a franchise since we get all this revenue from the phone company anyway?" Well, the city wasn't getting "all this" revenue! The city was not getting the revenue it thought it was. SWBT was short-changing them in a serious way.

Furthermore, the city revenues NCI would be providing did not come from the same sources as those being paid by SWBT. When customers used dedicated local loop access from NCI or NCI-provided Computer or LAN connection in the metro area these represented new net revenues for the City of Houston. It would be all new business.

This endeared NCI to George. It may have been the key in the negotiations to get the features NCI needed into the franchise. George never said that in so many words, but Scott believes it was part of the reason why he kept working with NCI even though he knew NCI was a startup, nothing more than two guys with ideas and no money. Of course, their experience in fiber optic networks also helped. They heavily leveraged Scott's fiber optic expertise to demonstrate NCI could get the job done. There was no more experienced company than NCI to put fiber in the streets and create a competitive local loop company. So who else could do it?

Another challenge involved getting adequate power to the equipment installed in buildings. Power was required to light up the fiber optic multiplexers. In 1986 no one had figured that out yet. The phone company used copper cable to carry the signal from CO to buildings. They did not have fiber optic multiplexer equipment in tenant buildings. They sent electrons for ring current power over the copper wires from the CO to the individual customers in buildings. This was not going to happen for fiber optics into a building. They needed power locally in each building for the fiber optic equipment. It seems like a trivial thing today, but then it was a huge hurdle.

NCI must rent service rooms in every building, route the fiber to the room and then install racks and equipment. Scott recruited one of the engineers, Gordon Schreiber,[43] who had an Apple Mac computer, to make a drawing to help explain it to the city as an example of how NCI was different from the phone company. It was a key concept to putting fibers in a building and became known generically as a building "Point of Presence."

---

[43] http://tinyurl.com/RoutingPaths-POP-Drawing

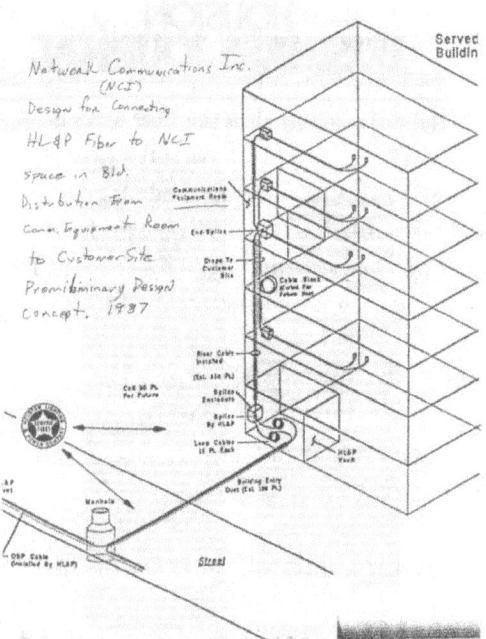

Later in 1987, once NCI had the franchise, this drawing[44] aided the effort again when he provided a copy to HL&P while negotiating the deal to use their conduits.

Metropolitan Fiber Systems (MFS) was apparently researching the same details by the time Scott met them in 1988. At the time, it was a hurdle to get into buildings to install and operate fiber optic equipment. NCI had to make it feasible to bring fiber systems into buildings. It was one more matter about which they had to convince the city they knew the business.

Scott convinced Susan Hodge and George White that he knew how to subcontract the work to experienced contractors and get it built. When they asked why, he explained that they already had experience with big installations for large Enterprise customers, substantial fiber systems at the time.

Scott recounted many examples in which he connected multiple buildings with fibers and used multiplexer equipment to carry voice, IBM data, LAN data, even analog video, over a shared fiber optic backbone that reached sometimes miles between buildings. They investigated NCI and learned they were leaders who had put in some very sophisticated fiber voice and data systems and made them work to support mission critical applications.

Even with confidence and experience, they faced hard questions about what they were doing.

They were verifying that they could:

1) Build a working network with fiber technology that originated in a customer's building and not in a Central Office (the phone company's model)
2) Legally and technologically provide the services envisioned
3) Obtain the right to enter buildings, dig in the streets, and use existing conduits
4) Make money selling local loop services in a metropolitan area (given that they were not the phone company)
5) Persuade customers to buy the networked services
6) Attract funding—if successful at obtaining everything else needed

---

[44] Note: This drawing, and several more documents from this era are PDF files that were created years ago by early PDF software. It has been called to our attention that some browser-based readers may have trouble with them, especially eReaders. Until we can update them, if you have trouble opening them please check that you have the latest browser updates, or try from another computer.

THE BLACK BOOK

The city had many questions that needed answering, too. Department staff had provided a questionnaire and NCI had to submit a "book" that provided the answers. Only then would they seriously consider awarding the franchise. Larry Koonce did a masterful job of responding to this.

Scott dubbed it the "Black Book" since it was a black binder. It was the official document that allowed NCI to fulfill the requirements necessary to become eligible for the franchise. Scott kept not only the Black Book itself, retrieved from the City once they'd passed through the gauntlet, but the very PC itself used to create the book. George White had given the original Black Book to Scott as a memento once it was no longer of use to the City. This was an unusual move, indicative of the respect Scott had earned among the City administration.

The black book was a compendium of all the plans and ideas with a lengthy explanation of the services NCI intended to provide to the City of Houston. It contained everything from a statement that set forth the benefits of allowing competition in local loop services to a bold (but well supported) prediction of growth in Local Area Networks and the demand for local loops.

It summarized three key service area offerings that would differentiate NCI from Southwestern Bell Telephone Company. Briefly, these included:

1) **Local Loop Circuits** (DS0, DS1 & DS3s) to connect end users (from their building) to the long-distance company, as well as to circuits between end user buildings and circuits between IXC sites. Putting fiber optics between the customer's building and the Inter-Exchange Carrier building provided immediate benefits in quality and redundancy because using fiber meant the service would not go down if a cable was cut or an electronic part failed (unlike the copper cable used by the phone company).

2) **Native Speed Ethernet** (10Mbps), or **Token Ring** (16Mbps), **LAN interconnection services** between end user buildings in a metropolitan area for Enterprise connectivity within the city. Other kinds of connections were also envisioned, such as RG62 coaxial connections for IBM terminals, or Dual Wang RG59 connections. Newly developing standards such as FDDI were also envisioned (even though FDDI was just a concept at the time).

3) **The flexibility to adapt to future services,** involving the use of ISDN (the use of ordinary telephone copper wires to transmit voice and data) for example, or cable TV service links. Since fiber to the home would require a citywide network, the network also could serve as the backbone for traffic carried to buildings or to neighborhoods someday. It could also accommodate the use of other technology, such as wireless if it ever evolved to become fast enough and reliable enough to go all over a metro area.

A copy of the application and business plan submitted to the city describes these early and groundbreaking plans for LAN connectivity. It's also noteworthy that the first part of the plan was printed on a dot-matrix printer and that this section was printed in October 1986. The rest was typed in the neat and clean font of an IBM Selectric.

NCI told people in 1985 and early 1986 that no one else had attempted to build such a metropolitan network. But by late 1986 news about other similar efforts began trickling out from a few other places around the country. Networks were underway, or already functioning, in New York City, Chicago, and Washington, D.C.

But there were critical differences between NCI and those other players.

Ron Vidal, who worked for a company called *Peter Kiewit Sons Inc.* (PKS), called Scott in the summer or fall of 1986. Scott was startled, and then worried, when he realized Ron knew about NCI and their local loop network plans. But then Larry set Scott at ease. He had met Ron at a seminar sponsored by Andy Lipman (a regulatory attorney) in 1986, well after NCI had applied to the city for the franchise.

There he also met Tony Pompliano and John Lucas of *Chicago Fiber Optics*. They were going to pull fiber through old abandoned coal delivery tunnels running under the streets of Chicago. He also heard about Teleport Communications in New York and Scott Brodie of ICC in Washington, DC.

None of these other concerns had gone after citywide franchises to build a fiber network. They were instead negotiating rights of way to pull fiber or dig up the streets and install fiber optics for route specific metropolitan area networks.

But as he found out there were others interested in metro networks it reinforced the feeling that this was a viable concept. In addition, it was reassuring to learn there were deals like ICC's out there. ICC was using venture capital money and bridge financing provided by the fiber optic contractor and banks. The fact that Peter Kiewit Sons, Inc. was out there bidding on similar deals like also encouraged NCI to press on. They were confident they could get someone to build the network. And the financing might come through Morrison-Knudson Corp.,[45] which had helped ICC in Washington, DC.

It was great to get confirmation that it was not such a crazy idea after all! And once he discovered Teleport, ICC, and *Chicago Fiber Optics* were out there, too, he made sure to mention them in the document submitted to the City of Houston as part of the Black Book. Their existence only further justified awarding the franchise.

In ongoing conversations with George and Susan, Scott emphasized that other cities were starting to build networks. The message was obvious. Houston had better get moving on the franchise if it hoped to attract the blooming number of large corporations that valued high-speed fiber-optic networks.

That message was developed into an overarching strategy to tie the city's ongoing civic push for "economic development" to the notion of fiber optic networks. High-speed data networks, they kept repeating, would demonstrably lift Houston's position as a city of the future and the "information age." They used LAN connectivity as an example of what could be done using fiber. Southwestern Bell had no such service offering. And while long-distance companies were beginning to lay fiber, that was not relevant to what Houston needed.

---

[45] https://en.wikipedia.org/wiki/Morrison-Knudsen

NCI stressed that when planning a metro network what needed doing was more in keeping with what they had done rather than what others might be doing to build fiber between cities. Scott pointed out that the infrastructure inside of a building between the floors, where fiber was used to interconnect voice, video and data systems, was the real long-term key to success for a metro fiber business that might locate in Houston. YSA had more experience at doing that than the phone company, or any long-distance company.

Southwestern Bell Telephone Company was selling dark fiber to customers in an attempt to kill the DS3 market before it developed. Bob Atkinson at *Teleport Communications Group* (TCG)[46] used to say that in Manhattan, Teleport could get $8,000 a month per customer for DS3 and $900 a month for a T1. We were hoping to get $3000 a month for DS3 and $500 for a T1 in Houston.

Larry and Scott, of course, were all the while discussing what would make the franchise deal viable for us and what would kill it. Any agreement would first have to be realistic and cover all the things the City of Houston routinely included in its franchise agreements. At some point, he said, "Let's just write one up based on the other ones that exist and give it to George and see what happens.

---

[46] https://en.wikipedia.org/wiki/Teleport_Communications_Group

Drafting the Agreement

Thus, NCI drafted a franchise agreement the city would later use as a model for others to follow. Larry Koonce wrote most of it after studying the existing templates of Houston's franchise agreements with the phone company, cable TV company, and HL&P power company. Scott defined the scope of services, the dry and technical part of the document. They sought to paint with as broad a brushstroke as possible. NCI construed their scope to cover "any communications service ..." in order to permit LAN connectivity. But also, because they did not want to preclude themselves in the future from making broad use of technology, including technology yet to be invented. The future of fiber optics was bright.

Larry Koonce sometimes talked about Scott's "Star Wars" ideas. He even had a stamp made that said, "Fiber Optics: Ra-Ra-Ra" and gleefully stamped those words all over the draft agreement documents. That enthusiasm, however, sure helped when it came time to telling the fiber optics story to city officials and potential investors.

Defining a "serving area" within the scope of services was crucial, of course. NCI certainly did not want to be limited geographically. Thus, the agreement defined an area of several hundred square miles that served more than three million people in 1986. The rights, as described, included fiber optics services supplied to homes. The inclusion of the use of the wording "any technology" assumed no restrictions would exist to providing fiber, copper or wireless service to offices or homes at any time in the future. This broad inclusion provided flexibility to evolve with technology and use more than one type of technology.

The agreement would create what was believed at the time to be the worlds' first metro fiber optic franchise covering the largest single metropolitan area in a major metropolitan city. Houston was the fourth largest city. One could drive for over an hour on the freeway and not leave the city limits. There was no similar single "serving area" larger than the City of Houston's to be found anywhere in the nation.

There are two key pages of the franchise that defined services as "any" service. NCI did not want to be limited to just telephony or data. Scott also envisioned carrying cable TV someday. This was not on his roadmap, rather merely a place-holder that could have long-term implications if the fiber ever went to homes. He knew fiber to the home could happen someday and delivering all kinds of services over fiber was a reasonable projection. He did not want the cable company to block them any more than the phone company.

They wrote it so they could use any technology to create any service and fervently hoped the City staff would agree. Even the New York City and Washington, D.C. networks were not built on a citywide franchise. They were doing voice local loop connectivity to long-distance carriers — not the LAN services NCI focused on. Neither Teleport nor ICC envisioned connecting LANs at native speeds as part of the business of an alternative access provider.

It was a bold and favorable agreement. The cable TV franchises, after all, had geographic limitations imposed on them and they were limited to selling only cable TV services. The phone company was limited to phone services. And the power company had a franchise for power services only.

Scott gives Larry Koonce credit for successfully negotiating this monumental agreement. He was brilliant at writing it up and brilliant at persuading the city to agree to retain the key ideas in the final draft version. Although they made changes, even much of the revised language came from the version Larry had prepared.

By late 1986 George and Susan had been won over. The city franchise department was going to send the franchise to the city council for approval. Scott and Larry were ecstatic. The awarding of Houston's cable TV franchises had been very political. Only the power players in Houston could obtain a franchise and it was divided among them so each one got a part of the city. NCI, on the other hand, was going for a citywide franchise with no geographic boundaries or service limitations.

They had no lobbyist, no attorneys, and no influential people working for them through this part of the process. They did not even buy Susan and George so much as a cup of coffee. It is still a little surprising to imagine they could work alone with these civil servants in an atmosphere of mutual trust and respect. Would this have been possible in another city? It seems doubtful. It's a testimony to George and Susan, and to former Mayor Kathy Whitmire, that they were even given the opportunity to present their case.

The city also wanted assurances that customers would buy the services. So NCI sent letters to Claydesta Communications[47] (later bought by Wiltel) and Texas Commerce Bank urging them to sign up quickly. They also sent a letter to HL&P to again try to get them to grant use their conduits.

During this time, Scott also took steps to reach out to the City of Austin, which had expressed an interest in a franchise for a competitive fiber optic network. They wanted to build in Austin and Dallas as soon as possible. NCI had to provide pricing and product plans to the city to show they had thought things through.

---

[47] https://en.wikipedia.org/wiki/Clayton_Williams

Waiting to be Crushed

Scott and Larry went through all of this in 1986 with nerves on edge. They were worried that SWBT would find out about NCI and crush them with their political clout even before it got to a city vote.

Business is, as much as anything, about relationships. Fortunately, their relationship with city staff was so good they began coaching us through the process. They had, of course, pointed the staff's attention to the fees SWBT should have been paying but wasn't. It seemed to help their case further that SWBT, up for renewal of its own franchise, was being cocky and arrogant. Southwestern Bell acted as if it could run its business without a city franchise and the city could do little to stop it. After meetings in which Bell treated the city franchise department in this shabby manner, the franchise people would meet with NCI. these meetings were purely business, too, centered on the franchise efforts. Among the key issues they needed information on, was the assertion that there was a need for competition.

They were receptive. Relationships nourish business, and this shows precisely why and how.

I believe Bell helped them in this regard by being arrogant. Larry Koonce would say that no one who then worked at SWB or HL&P had ever had to negotiate an initial franchise. They were just granted one. This illustrates one reason why monopolies always falter in the face of competition.

NCI, on the other hand, had to create the franchise and get the city to want to give it to NCI and not someone else. They had to further convince city staff to put their names on the line in front of the mayor and city council and recommend the city grant the franchise. And not just any franchise -- a new kind of franchise that did not exist anywhere else in late 1986 and early 1987. It was a big show of faith. The staff had to check NCI out thoroughly. They grilled Scott and Larry enough apparently to finally believe in them.

It took until early 1987, but they were finally put on the city council agenda. Although they'd been working on the application for more than a year, it seemed hard to believe they'd finally made it this far.

Scott and Larry were now doing everything possible to line up key people to work for them. Ron Vidal was bringing in key players inside Kiewit to work with us to bid on the build package they were putting together. Meanwhile, they were finding contractors to send bids that could compete with Kiewit. Scott was also out finding customers without drawing too much attention. NCI's advantage was the sneak attack strategy. SWBT couldn't exercise political clout against something it didn't know existed or see coming.

Larry and Scott decided it was time to check out what happened at city council meetings. Larry went down one day and sat in a pew in the middle of the council's chamber. He was striving to be inconspicuous. As he sat there several students from Texas Southern University entered and sat in front of him waving signs of protest. Shortly, more students sat down behind him. Before he knew it, he was in the middle of fifty black college students who were now chanting, too. Reporters and TV cameras appeared. And there was Larry Koonce, a non-student in a suit, right in the middle as the TV cameras and bright lights flicked on. They had a good laugh seeing him later, in the news broadcasts.

Feeling some confidence that they would get the franchise, they also began exploring raising money. Scott called his uncle John Glenn Yeager for advice. He had made a lot of money on his own when he left Exxon and he had the reputation of being the one in the family that had done well financially and understood business issues. Scott explained what they were doing and he became very interested in the deal. Later he would also invest in it.

The franchise was put on the agenda and they were told that the franchise had to be read at the council meeting and voted on three times to win formal approval. The same day and meeting at which the NCI franchise was up for a first vote, the franchise renewal sought by SWBT was on the agenda for a second reading.

When they saw the agenda, they anticipated a response was forthcoming from SWBT. But one didn't come. Perhaps Southwest Bell did not know about NCI until this second reading of their franchise. Whatever the case, NCI passed the vote on first reading. And Southwest passed its second reading. At the next meeting, Southwest Bell met with approval on the third reading of their franchise renewal and NCI passed their second reading. It was curious. They were meeting no objections from the telephone company. They wondered if perhaps SWBT was staying silent only until it was assured of its renewal.

Once NCI knew they'd made the agenda, they were encouraged by George White and Susan Hodge to retain a lobbyist. They hired Clint Hackney and he paid a visit to each council member. He'd been an elected state representative from Houston and was known for his advocacy on issues that benefitted the city (when he agreed with them, of course). He knew the council members and mayor since he was a state representative for the region of the City of Houston and had lobbied for key initiatives the City of Houston had wanted to be passed by the state legislature.

They were very fortunate to be able to get him to work on the franchise. His personal relationship with the City of Houston and his lifetime friendship with Scott's wife Susan made a big difference in his willingness to help us. He also liked that they were a couple of small guys fighting a big monopoly.

Clint got them in front of city council members at the mayor's luncheon and encouraged them to construct a letter for city council members explaining the benefits of an NCI franchise.

"BAD FOR BELL"

NCI had already passed two readings of the franchise ordinance and was coming up for their third reading. They were conjecturing that being up for the previous readings at the same time as SWBT's franchise was up for a reading was helpful. How could they raise a stink until they had their franchise renewed? They also wondered if George and Susan knew that would be the case when they arranged for NCI to be up for approval at the same time as SWBT, at least for the first two votes.

Things changed dramatically on the way to the third reading. Right there in the council chambers, in front of everyone, one of the city council members received a phone call from an SWBT lobbyist during the meeting. The council member held up the phone and said so everyone could hear, "*I have the Bell rep on the phone. He says this franchise is bad for Southwestern Bell and for the City of Houston. I recommend we hold off on the third and final vote. We need to study this some more.*"

Their hearts sank. Their pocket books were getting lighter. They were very concerned. NCI didn't have political clout like SWBT. Some comments were made about Bell wanting to stop competition. But Bell had plenty of "chits." And they immediately used some to delay the vote. Now they had a real reason to fear SWBT would crush them.

SWBT were successful at delaying the third reading for two weeks. Larry and Scott went to meet city council members and the mayor. They gave them their pitch about fiber optics and the Information Age and why it was good for the city. When the reading came up they would know who they were and what they were about. They also had to convince people like Leonard Childress to support them. Leonard was the telecommunications director for the City of Houston. He had a degree in political science. He also was strong politically within the city structure. He said they were not crazy; his word carried weight. John Glenn Yeager also helped with Jim Greenwood and other council members he knew.

SWBT, meanwhile, sent out a poorly written letter to council members. It arrived at the end of the two-week delay on the February morning of the third reading. It looked as if it had been written on a typewriter that needed a new ribbon, the quality of the print so poor it was hard to read. It appeared they were not putting their best competitive spirit into the effort.

Ultimately, the contest came down to last minute hallway negotiations between the city attorney and the SWBT lobbyist. In the meantime, Scott and Larry were rushing around to visit city council members with Clint Hackney, whose great relationship with city council members gave NCI political credibility.

NCI had written their response to SWBT, emphasizing one point; how this franchise benefited the City of Houston. Clint, who had been instrumental in overcoming the SWBT letter of complaint, made sure the response was read by city council members and that they knew who NCI was.

The day of the council meetings and the third reading of the city franchise agreement, Scott had pneumonia, but he had to go to the city council meeting in case he was needed to deal with last minute problems. In addition to sending the letter stating their objection to their franchise, SWBT sent Council Member Jim Westmoreland to stop the deal. With the council agenda time slot unknown, Scott and Larry had to be present for the entire city council meeting.

A Desperate Sham

During a break, Scott went downstairs with Clint Hackney and stood by the Coke machine. Clint went into council chambers, meanwhile, to check on things and find out who was with and who was against NCI. He came back to report that Westmoreland was opposed. Just then Westmoreland himself came out and pulled Clint and Scott aside in the basement to make a deal.

Westmoreland said SWBT would agree to the franchise if they would modify a few sentences to prohibit NCI from selling dial tone. NCI already had a sentence stating it could not do anything without approval by the PUC, so this was a stall tactic. Scott said that if — in the interest of getting it done — they could get the franchise approved that day, NCI would allow the change.

That was very nearly their undoing. Fortunately, more experienced minds exposed the sham. City Attorney Clarence West, who had a reputation as someone not to be crossed, was brought into the discussion. Clint knew him well and had worked on many other city matters, and had briefed him earlier. The city franchise staff had also briefed him.

West, fortunately, noted that any change in the language would require three new readings — a disaster that would send NCI through the whole process again. He urged them to refuse Westmoreland's request. Scott felt he had the city attorney on his side: Clint took that information and went back to Westmoreland, saying NCI rejected the compromise.

SWBT had tried to trick the inexperienced newcomers. Had they succeeded they would have pulled out all the stops to defeat NCI, sending in their whole lobbying team if needed. But for a bit of timely advice, the franchise would never have happened!

NCI Wins!

When NCI's turn came on the agenda there was a discussion of the SWBT letter and concerns about digging up the street for the cables. A key person in the city's public utility department (who happened to know Larry Koonce's father from some social club) testified that the franchise would not result in extensive trenching of the streets. Raising this issue was a last-ditch effort by SWBT and it almost worked. Luck held and Larry's dad's friend deflected it smartly, adding a comment that dismissed SWBT and their letter very nicely.

Then the franchise was put up for a final vote, which passed.

Scott and Larry were ecstatic and stunned. Everything had worked out perfectly. They'd beaten a $7 billion company with plain old hustle and some good luck.

From NCI to MFS

The franchise was awarded to NCI in February 1987. In March, John Glenn Yeager decided to invest in NCI, taking a huge, early risk on NCI's success. His early investment was critical, providing the money they needed to live, pay the IRS, and make house payments while addressing the myriad details for the franchise. By April NCI was giving presentations to other potential investors, and anyone else who would listen. There was a good story to be told about this new industry and Scott was inspired to tell it.

NCI was also evolving the concept of *Alternative Access Vendors* (AAVs) with a handful of people lead by Andy Lipman,[48] a Washington, D.C.-based regulatory attorney. Many see him as a key force behind the rise of the alternative access provider business. Scott has maintained a close working relationship with him over the years.

The AAV notion predated Alternative Local Telephone Services (ALTS) and Competitive Access Providers (CAP). Andy is a great innovator and helped get the industry started.

After landing the Houston franchise NCI had to flesh out the business plan for the fiber network in preparation for the bid specifications they would soon be sharing with potential contractors. Scott had created projections for the city, but this was a much more detailed exercise.

To size the market, NCI hired college students to visit the 40 or so buildings planned for the downtown network. They established contact with building tenants. NCI estimated the market using a variety of approaches. In the end, they concluded the market would generate revenues of approximately $533,000 per month through sales of DS0, DS1, and DS3 circuits.

How accurate were those estimates? They predicted that by 1993 NCI would be showing revenues of $6 million annually, assuming by then the network would also include Houston. In 1993 MFS Houston would indeed show revenues of $500,000 per month, or $6 million annually.

---

[48] https://www.morganlewis.com/bios/andrewlipman

HOUSTON NETWORK ARCHITECTURE

The network architecture for Houston used revenue projections of $500,000 per month or $6 million annually, estimates later validated. It was a challenge, however, in developing these in such a way that they could manage and switch the transmission equipment down to the DS0 level. None of the industry players had the equipment to do that. So NCI turned to a Tulsa Power company engineer named Mark Keeter who had built a nonstandard network to be used for their internal utility network that was similar.

It was not the common architecture to have multiplexers integrated with digital switches. Plus, they were doing this in support of this totally new concept of a metropolitan fiber network. It took until 1996 for the industry to catch up to the NCI 1988 design. By 1996, add-drop and insert multiplexers with Digital Cross-connects as a type of network which used SONET and distributed 0-1-3-DACs equipment became the industry standard. It appears their approach was ahead of its time. The singular Houston design of 1988 became the model by which all metro networks functioned.

This 1988 design also allowed NCI to offer high-speed data services such as Ethernet and Token Ring. Some day they said they would be able to do FDDI services over the same fiber plant. SONET was not yet the approved standard. AT&T had a standard they were pushing in lieu of SONET. It was asynchronous and no one knew if that standard would beat out SONET in the late 1980s and early 1990s. The choice was between proven big name companies or choosing a new standard for network approach.

Another idea included in NCI's architecture provided protected or unprotected service for T1s. They were thinking in terms of charging a premium for self-healing functionality. Bell was not protected. They were on copper, not fiber. NCI would just use half the ring if the customer wanted the service cheaper than Bell could provide, or provide at premium cost a service Bell could not match.

NCI developed a specification package in 1987 and put the work out for bid to Peter Kiewit Sons Inc., Siecor Corporation, FiberLAN Inc., Gordon Schreiber, Morrison-Knudson Corporation, and Northern Telecom Inc. The Siecor/NEC bid to network the 34 buildings in downtown Houston came in within the $7 million range, Gordon Schreiber was at a little over $6 million, and FiberLAN was at $8 million. Peter Kiewit Sons, Inc. had a close relationship with AT&T. So they returned a bid that used the AT&T standard with a self-healing ring and add-drop capabilities. It was more expensive than back-to-back multiplexers with a matrix switch between them. The PKS bid topped $12 million.

As already noted, the bid specifications were a bit unusual versus the approach accepted using proven long haul fiber optic transmission equipment. There were a few heated discussions with Jim Crowe of PKS on the subject during the bidder conferences. He did not agree with the network architecture since it was unproven. He thought it too risky. At the time Scott disagreed. He realized later that Crowe was right when it came to the risk new technology can pose to a startup company.

NCI was to deliver services to customers who were not forgiving and could easily have acquired a bad reputation if a new technology failed. This would be the first of several times Scott would disagree with Jim Crowe and he would turn out to be correct for that decision in that context.

The record shows that NCI's envisioned approach of protecting these services in 1987-88 was not the direction the industry wanted at that time. However, SONET did become the norm by 1992 and the functionality NCI had envisioned did become reality. After working with equipment and technologies that are new and sometimes fail to live up to expectations, it is clear in hindsight that Crowe was correct to demand off-the-shelf standard technology.

The specifications NCI put out for bid were different than what people who built fiber networks were used to seeing. Scott felt from the beginning that multiple services over the fiber network were necessary to get the revenue needed to run a profitable network. The DS0, DS1 and DS3 business was simply not sufficient. Scott had expected from the beginning to incorporate data services to increase the total revenue.

In terms of architecture NCI also must penetrate the power company entrance facility, known as a vault, where they enter a high-rise building. Also, NCI was developing their own equipment rooms in the targeted buildings. Bringing the fiber optic cable into the building via the Vault was essential to getting access with HL&P cooperation. The building *Point of Presence* (PoP) was also essential to providing the necessary power to the fiber optic equipment. Another reason for creating a PoP in a building was to create a central point from which to distribute cables to tenants in the building.

Scott had explained all this to city officials to obtain the franchise in 1986/87. He had to explain it again to HL&P to gain their cooperation and access to the vault. Gordon Schreiber's Apple drawn image[49] for the plan proved critical several times and that drawing ultimately defined the standard way all metro fiber networks would build fiber to large buildings.

Houston was meanwhile pursuing a private fiber optic network bid. The city wanted to link its police and fire stations together including the Motorola radios they used for mobile communications, a leading-edge solution as well. City employees in the Telecommunications Department knew Scott from his franchise efforts. Thus, he sent a letter to the city inquiring about bidding on the network. L.C. Baird with the fiber optic construction arm of Peter Kiewit and Sons worked with NCI to submit a bid.

This project kept the ties close between Kiewit and YSA, Scott's fiber optic systems integration company. Scott favored NCI using another contractor to build the franchise network but still collaborating with LC Baird/Kiewit on the City of Houston deal. Scott thought keeping open a relationship with Kiewit was a good idea since they were big and he could perhaps help them get the city's private project independently.

---

[49] http://tinyurl.com/RoutingPaths-POP-Drawing

In May 1988 NCI had the project financing lined up through First Interstate Bank, the bank that had a Letter of Credit from Siecor the fiber optic cable manufacturer for the $7 million in construction financing. NCI selected FiberLAN as the construction company for several reasons. They believed in the concept of using fiber to deliver more than North American Hierarchy Telephony circuits, for one thing. Scott believed LAN connectivity functionality was a key differentiator that allowed NCI to offer services the customers wanted and had to be reflected in the design.

Black Monday

NCI still needed to arrange $3 million in equity investment to leverage the $7 million in debt. It was an attractive package to investors because the risk above the $3 million was not on the investors but the equipment manufacturer. This was a good deal for those investors because they were putting up $3 million on a $10 million risk.

Steve McVoy and his friend Bill Mack (grandson of Senator Connie Mack of Florida) were in for $1.5 million. NCI was talking to Gerald D. Hines, a national real-estate developer based in Houston. The industry was moving slowly, and they were looking at other alternative deals as well.

Crowe contacted NCI in late 1987 to see if they might make a deal. Larry thought NCI could raise the money via other sources and did not want to negotiate seriously. Scott was ready to move. NCI stuck it out and tried to raise money via FiberLAN and other means. Later money opportunities looked good but never developed

Ned Holmes, who was at the time chairman and president of Parkway Investments and Texas Inc., had just agreed to fund NCI.

A month later, on October 21, 1987, Scott was sitting in the office with Ned Holmes and watched his facial expression suddenly change when someone entered the conference room and handed him a note. It was "Black Monday," the day when stock markets around the world crashed. The DJIA dropped 508 points, 22.6 percent in value. Ned left the room and when he returned he said he had just lost a lot of money. Scott, who didn't own stock at that time, didn't comprehend the experience of making or losing money in the stock market, but he did recognize that this meant they were not going to get the money from Ned.

Ned attempted to renegotiate their deal with unfavorable terms, and NCI finally said "no" after spending over $20,000 in legal fees they did not have to waste. Jack Clark handled work on credit. Had he not, NCI would have been in serious trouble.

Peter Kiewit Sons' Inc. return

As NCI continued hunting for financing, the most promising financial arrangements were still with Peter Kiewit Sons' Inc. When they asked for a second meeting, Scott thought it was to make a deal in which they would invest in NCI and they would be taken over by them. They had had similar talks a year earlier in 1987.

Larry appeared at the second meeting in 1988 in no mood to negotiate. Not only did he not cooperate he made no effort to present anything reasonable to Kiewit. Scott was beyond upset. He didn't agree with Larry, who felt he had the financing and most of the money lined up and thus had no need for Kiewit. Scott was eager to get moving, to start building the network. And Kiewit could make that happen. Larry, meanwhile, was still exploring financing options.

Crowe and Royce Holland, also of Kiewit, were gracious to Scott, but Larry was pursuing another agenda. He had not let Scott know about his thinking ahead of the meeting. The meeting was a disaster. The meeting ended on such a bad note Scott thought the relationship with Kiewit was finished. It was not a good time to lose this partnership. Investors were hard to find.

 For example, a letter in the files dated December of that year from Gerald D. Hines[50] illustrated the attitude of investors. It speaks especially clearly to their fears about the monopoly in each city, the RBOCs and how risky this new network building business was in their view. The entire letter may be read on the website, but excerpting a few quotes serves to illustrate their concerns:

Will Southwestern Bell be successful in their apparent efforts to pre-empt competition in the local loop? Both S.R.23.27 and ICB's are means by which SWB could engage in predatory behavior.

Is the current market large enough and can NCI's market penetration be high enough to reach breakeven ever?

Are the economics of installing such an extensive physical network sufficient barriers to entry to keep another player from entering the market?

---

[50] Gerald D. Hines Interests Letter Dated December 28, 1988
http://tinyurl.com/RoutingPaths-HinesLetter

DARK FIBER THREAT

In sharp contrast to the view of investors like Gerald D. Hines there was the optimistic view of Crowe and Royce Holland. They did not quail at the thought of competing with the RBOCs. Royce Holland's steady temperament was especially instructive. Scott, on the other hand, let the SWBT monopoly rattle him, especially when Baby Bell minions falsely claimed they would somehow put them out of business. Later, in the 1990s he was fortunate to learn still more first hand from Royce Holland in the regulatory arena. Royce and Andy Lipman were not only very calm but also very clear in their explanations, and these two attributes accounted for how MFS Communications could make so much steady regulatory progress. MFS increased its valuation over a matter of a few years, going public with a $1.4 billion valuation in 1993 and selling for $14.4 billion[51] to WorldCom three years later.

The regulatory battles NCI fought in 1987 and 1988 also required them to deliver clear explanations of new and often difficult concepts to help regulators understand the issues quickly and in non-technical terms. The main battle at that time was fought over the concept of Bell selling dark fiber and using *Individual Case Basis* (ICB) pricing to reduce the price customers paid for NCI's biggest revenue opportunities for DS3s. They were only priced low to the biggest customers.

Dark fiber is essentially unused fiber optic cable that is available for future expansion. Another use of Dark Fiber was to lease a big customer dark fiber so they did not need big circuits such as DS3s on fiber.

The owner of the cable was a utility like SWBT. Only companies who were utilities or special fiber carriers like NCI could install fiber in the city. NCI had spent over a year and taken a huge risk to obtain a franchise. NCI could either dig up the streets directly, or make a deal with SWBT or the Power Utility Company HL&P for the use of their conduits. These agreements were tough to get.

SWBT would not lease NCI dark fiber to build a competitive fiber network. When SWBT did provide dark fiber, it was to a major company who was a customer for many other services, but not to a potential competitor. This experience was a harbinger to the local copper-loop deregulation in 1996, showing the need for competitive access. There was no competitive access in 1988, indeed the idea was considered laughable.

Dark fiber networks did not exist but SWBT preemptively created special custom deals to stop competitors from getting orders. SWBT was doing customer ICB deals for dark fiber at that time to allow only big customers to have use of their rights of way without actually getting a franchise. This was a very clever way of getting around their franchise fees and stopping competition from getting a foothold in the market. By using assets in the rate base that were subsidized by the *Plain Old Telephone Services* (POTS) infrastructure and guaranteed a rate of return in the rate base, they could subsidize such predatory behavior. NCI had no guaranteed rate of return and the investors had to take risks to build the fiber. Scott had taken huge risks to get the franchise, and had no rate base to guarantee any return.

---

[51] WorldCom Reaches Pact to Buy MFS in $14.4 Billion Stock Deal
http://www.wsj.com/articles/SB841024262424406500

This blatant anti-competitive behavior of SWBT was exactly what made the Gerald D. Hines people afraid to invest. They knew the competition had billions in revenues, a guaranteed rate of return on their investment in fiber, copper and switches and could lose hundreds of millions putting NCI out of business. That was exactly what the Dark fiber ICB effort was all about.

While these skirmishes with SWBT were demanding attention NCI was also worrying about financing. They had yet to bring on a lead investor. NCI was watching helplessly, worrying about attracting investments even while watching SWBT execute this very smart play that wiped out any early gravy revenue they might have expected from big customers for DS3s.

They could see that these large Houston DS3 opportunities with the IXCs were the very sources of large customer revenues fellow network builders were already enjoying in Chicago, Washington D.C. and New York City. Of course, this meant that Chicago was up and running, Washington D.C. was moving forward with ICC, and New York City was going forward with Teleport. This suggested that where once NCI may have been out in front they were now beginning to lag other markets. The word also was out that they had the franchise but lacked the money to build the network.

The SWBT attacks were part of their deregulation filing and began in 1987, just a few months after they received the franchise. The result was a regulatory rule-making process called SR23.27 that required NCI to file comments at the PUC in Austin Texas.

 In the end, the FCC changed its mind on DS3, ICBs, and dark fiber. This was partially because of NCI meeting and filing of comments with the FCC (Docket #88-136)[52] in November 1988. These filings got their attention, along with the efforts of ALTS and, later, Andy Lipman who was working for MFS.

Staffers at the FCC and regulatory people at AT&T and MCI said Scott's in-person and written comments had a huge impact. It impressed them that a small entrepreneur with no legal help and no lobbyist, used logic and direct research to effectively demonstrate the cross-subsidization and anti-competitive practices of the RBOCs, especially SWBT.

It was a simple beer analogy that grabbed them. The cost of a truckload of beer should not be lower than a single beer, he said. Everyone can relate to the fact that a can of beer is cheaper to buy when it is in a case than when a single beer is purchased out of the cooler at a convenience store and sold as a single beer. And as everyone who was in college had bought an entire case of beer (24 DS0 = T1), it was a wild fantasy that someone could buy an entire truckload of beer (DS3). No one would ever consider that you could buy a truck load of beer for one dollar or the price of a single beer at the convenience store.

---

[52] http://tinyurl.com/RoutingPaths-FCC-88-136

The ICB tariff which allows a big customer to buy a DS3 or "truck load of beer" for less than "the price of a single beer" was obviously a form of anti-competitive cross subsidization behavior, yet that was exactly what Southwestern Bell was doing. Using empirical data uncovered by digging into the FCC files on his own, Scott showed precisely what SWBT was doing with its sales of DS0 (single beers), DS1/T1 (cases of beer) and DS3 (truckloads of beer) circuits.

Scott had no staff and had no accounting department. He also had no secretary and no legal team. So, in 1988, the filing[53] was done on an IBM PC clone with Lotus 123 spreadsheet software. Lotus 123 and a PC gave him the equivalent of an accounting department with analytic tools and a staff of researchers with a mainframe computer. He must figure out what was required, manually enter it into a PC spreadsheet and produce a report that would have previously taken an army of people.

At that time, letter-quality printers were rare and expensive. The cover letter was professionally typed on a business typewriter and the spreadsheet report was printed on a dot-matrix printer, illustrating the rapid technological evolution since 1988. The FCC was unused to one person doing all the work, filing the document and then coming and presenting the argument in person.

There is an important point here. The RBOC had armies of people producing such reports and applications. The very complexity of a filing partly existed to intimidate and inhibit upstart, small operations such as NCI. Until technology put inexpensive computing resources in the hands of the small businessman, this was the exclusive playground of the huge bureaucracy. Commodity computing was unleashing a revolution that would completely change the business landscape.

Given how precarious the whole Enterprise felt these were very big victories to influence anything at the PUC or the FCC. They were important to the potential customers and investors.

They were important to everyone involved with NCI.

NCI had no money. Bell was trying to crush them. And Scott's other business, YSA, was under strain because he did not devote the time and attention necessary. He was still YSA's main sales person and sold a lot of product, so his absence had a profound impact on revenues.

Still, YSA managed to move forward. Between 1985 and 1986, YSA designed a new TDM multiplexer for IBM system 34/36 computers working with a Honeywell engineer. YSA installed the first ever fiber optic system for this family of IBM computers. Significantly, with fiber-optic cable runs of more than 5,000-feet at the Rohm and Haas plant in Houston.

IBM doubted it could be done, YSA proved them wrong by doing it.

They also installed a fiber system for Gould Modicum Programmable Controllers in an HL&P power plant in Jewitt, Texas. This was the first turnkey project they ever delivered as YSA. It was also the first time anyone used fiber on that brand of programmable controllers for their mission critical Power Plant Control Systems.

---

[53] http://tinyurl.com/RoutingPaths-NCI-FCC-Filing

And, of course, it had to work. The customer would plug in their equipment on both ends of the fiber optic cables and multiplexers and if the system did not work they would know it immediately. YSA had to get the orders, install the systems and count on everything working as a system. If it failed, these projects could have pushed YSA into bankruptcy. So of course, failure was not an option. By 1987 Scott sold out to Philip Sumners his brother in law and had no income coming from YSA. His wife Susan was generating the only revenue the family had to live on from 1987 to 1989 other than the small about paid to him when his Uncle invested in NCI.

It was quite the roller coaster ride. They were on the leading edge of the opening of the metropolitan local fiber optic network business. The telephone long-distance and PBX telephone equipment businesses were less than 10 years old and satellite communications markets were brand new. Cell phone networks and Wi-Fi did not exist. There was no Commercial Internet. It wasn't even a germ of an idea, yet.

Scott was worried they were being unrealistic, attempting a deal on this scale with no funding in hand. No one appeared to have managed this before. The deal that Scott Brodie had done in D.C. maybe came closest.

Then he met Barbara Sampson, the founder of *Intermedia Communications, Inc.* (ICI)[54] of Florida. She was getting started on a fiber network in the Tampa area. They discussed the issues and the market and he talked network architecture with her technical people. Soon he was collaborating with her, as well as with Bob Atkinson of Teleport Communications in New York. Later, Larry and Scott met with Tony Pompliano and John Lucas from *Chicago Fiber Optics*, who asked if they were interested in becoming part of their deal. Scott was becoming visible on the national regulatory scene, which led to NCI joining with Andy Lipman and Bob Atkinson in forming the *Association for Local Telephone Services* (ALTS) as a charter member.

The aim of ALTS, as set forth in its charter, says a lot about where this totally new industry was headed. The charter said their immediate aim was to "establish and maintain a legal and regulatory environment at federal, state and local levels that supports the development of competitive local telecommunications transmission companies and to foster the exchange of ideas between existing and potential local telephone transmission companies."

Those filings with the Texas PUC and the FCC demonstrated that Scott, as NCI, was doing something new and fighting the phone company even though newcomers, although they did have the first Citywide franchise. He had explained those few examples of how dark fiber was bad business for the rate payer who funded the phone company via the rate base. That really opened up the FCC and PUC staff members' eyes. He also emphasized the value of the new high-speed data services NCI was offering. He used a Token Ring example to show how NCI could offer services the phone company did not. He wanted them to understand that there was more than competition for existing services at stake. These new high-speed LAN services were going to be how companies interconnected their buildings and conducted business in a networked manner.

Clearly, there was a lot going on in the NCI business world in 1987 and 1988, but nothing quite the equal of the change that was about to occur.

---

[54] https://en.wikipedia.org/wiki/MCI_Inc.

MFS Buys NCI

Metropolitan Fiber Systems (MFS) was created in 1988 in Chicago under the umbrella of Peter Kiewit Sons Inc. (PKS) and, like NCI, it set about providing telephony connections to businesses through fiber, though data was not on their minds then. Their fiber cable runs mostly passed through underground freight tunnels left over from the turn of the Century in downtown Chicago. The tunnels were built in the 19th century for use by small freight trains carrying packages, trash, and coal underneath the streets. They also provided access for telegraph, Teletype, and telephone cables.

PKS is a major construction contractor associated with major projects such as Boston's Big Dig and California's Devil's Slide Tunnel. They were not in the telephony business and did not seek to be in that business, though they did lay cables and other construction related to the business. On March 12, 1986, *Chicago Fiber Optics Corporation* entered into a contract with the City of Chicago, agreeing to pay a franchise fee and a percent of gross billings for the right to provide services via the tunnels. PKS contracted to install a series of fiber optic cables into the tunnels for the company. Under-capitalized, *Chicago Fiber Optics* floundered and became unable to pay their bills. PKS assumed control of the company in lieu of payment and in 1988 began operating the service. After deciding to expand the business model to other cities, the name was changed to *Metropolitan Fiber Systems*, and expansion to several additional cities initiated.

Thus, it was a natural thing for MFS to buy companies that had inroads to target cities. NCI had a newly minted franchise in Houston, but no capital and no operational business. NCI was perfect for MFS.

The MFS Deal

NCI intended to make use of utility rights-of-way. It was a brand-new business model company. There was no last mile long-distance local loop provider the city and the commercial Internet was not available yet. These newly created services by NCI, Chicago Fiber Optics, ICC and Teleport Communications were the first ones to provide fiber rings in major central business districts.

Larry wanted out of the business and left. Scott and his Uncle John took over NCI to keep it running. Scott brought experience, knowledge and sales ability to the table. John bought management skills and his investment capital.

Jim Crowe and Royce Holland of Kiewit, who had talked with NCI twice about the business, were the top executives of MFS. Crowe was Chairman of the Board and had the ear of Walter Scott, CEO of PKS. Royce Holland was CEO and later became the president. From the time they launched, MFS knew NCI.

John and Scott then went to Crowe in Omaha and made a deal. The deal was contingent on Scott getting some changes in the arrangement with HL&P for the use of their conduits. These agreements were tough to get.

Scott and John signed the deal with MFS and set about getting the rights of way deal modified with HL&P to get the network under construction.

Teleport Horns In

Just as they were negotiating with Crowe, Teleport Communications came to town. Scott had helped them earlier because he trusted them, now they took his information and expedited their own ability to get a franchise in Houston. He had been naïve to have trusted them and this was his reward.

NCI fought them and lobbied but they did not have to mount the same set of hurdles that NCI had surmounted. The City Council of Houston granted Teleport Communications a franchise with zero hassle. It became obvious that MFS also could get a franchise in Houston without NCI. The value of NCI's franchise was diminished, but Crowe had said he wanted Scott as well as the franchise. NCI needed the money and although they had them on the ropes, MFS honored the deal. They could have backed out at the last minute with impunity, but Scott recognized that these were fundamentally honorable people. Scott also recognized that MFS was ahead in planning and construction expertise. The handshake deal with Crowe occurred in February 1989, a full two years after NCI had won the franchise.

Employed by MFS

Fortunately, Crowe also agreed to hire Scott as a consultant during the renegotiation period with HL&P. He allowed Scott to invoice Kiewit for the months of March through July. Then he became an MFS employee.

Scott was interviewed by Tony Pompliano, who became a company president. Tony introduced him to another employee at that first meeting. Scott expressed shock at Tony's abusive behavior toward the man. This was a harbinger of future abuses, as this was Tony's way of leading his key guys. Scott did not like that and found the man to be a very hard worker, undeserving of such abuse.

Scott respected what Tony Pompliano pulled off to get Chicago Fiber Optics going and how he parlayed that into this position at MFS.

The Challenge of Market Sizing

Not long after NCI sized the market in Houston with the grassroots market study, they learned that they were right about the number of DS0, DS1 and DS3 circuits needed by the 43 buildings targeted in downtown Houston. Scott later learned that Barbara Sampson at ICI sized the market in Tampa for T1s at five times Houston's! And when he shared his numbers with the people at the investment banking firm of Allen Patricoff and Associates, they insisted he had sized the market too small.

Gary Lasher, the CEO of another CLEC in Philadelphia also was wrong about the size of the market. He was going up against MFS out of the chute.

The model Scott had developed to launch NCI was the correct one, multiple services were needed to justify building a network in most cities, just as he had said would be the case in Houston. DS0, DS1 and DS3 connections alone did not generate enough revenue, but tremendous resistance had to be overcome to gain those additional services.

NCI was right about the model and yet did not get the money while others were wrong but still got the money. The lesson learned: building a business is not the same as building value to raise money, or building perceived value to go public and become a hot stock. Conflicting issues of realistic projections and optimistic, or overly optimistic projections require a difficult balance.

Investors and the stock market demand wildly optimistic growth curves. Without extraordinary expectations, they will never invest. Their demands are not reality-based. A plan may have a great growth curve and project reasonable, achievable growth in revenues, yet that is not sufficient. Reasonableness means the "money people" won't back you. Luckily, Kiewit was privately funded and had the money to launch and set MFS to building the physical network without having to convince investors. It had both the vision and the deep pockets to see it through.

Scott ended up at the best place possible by going with Kiewit and MFS. Not only did he have a job, he also had stock in MFS of Houston Inc. from the sale of NCI.

Pulling Fiber in Houston

Scott was MFS City Director for Houston, charged with getting the network built, coordinating building access (which included getting the right to pull fiber in a building and to own a PoP), acquiring phones and furniture, hiring a sales force, growing revenues, and doing whatever was needed to build the business. He also had to coordinate with operations and installation crews when installing an order to make sure the customers were happy.

He laid out the fiber to the key buildings and decided which buildings to connect with fiber optics. He had performed the original market study in pursuit of the NCI franchise and had negotiated a deal the power company. He also had sold the idea of a PoP to get fiber into a building and install the equipment.

Scott was ready to move fast.

Scott worked with the construction people from Kiewit to see the fiber pulled to the major buildings in town. This required constant coordination with the building owners and PKS construction people involved. And, of course, they had to reach all the buildings targeted. Scott diligently saved documents and materials, and carefully documented every step in those first years. His habit of maintaining files thoroughly documents the history of his projects, and populate his website.[55] The cache includes, for example, an August 1989 article[56] in the *Houston Business Journal* that carried news of the MFS acquisition of NCI.

---

[55] http://www.fscottyeager.com/
[56] 1989 Houston Business Journal NCI to MFS
        http://tinyurl.com/RoutingPaths-HBJ-NCI-MFS

Building the "Node"

In addition to a PoP in each building, a central point where all the fibers terminated is necessary. Much like the phone company had a CO where the copper loops terminated, the fiber-optic network needed a central aggregation point. This was designated the "Node" in MFS nomenclature, though it served precisely the same purpose as a CO.

Scott and his MFS team had to provide details that helped tell the story to potential customers, building owners and HL&P executives. Many photos are on the website, including shots of the Kiewit people pulling fiber. They were amazingly skilled at getting fiber and multiplexers installed and guided HL&P as it pulled the fiber so MFS could then build out the node. Also included are photos of the node itself, the central location all the fiber terminated and tied into the TDM multiplexers. This is where the lasers put modulated light into the fibers, creating such service offerings as DS3s and DS1s.

The story included the floor plan for the office space, the node room, and the placement of the multiplexers (MUX). They developed an overview spec sheet of the products, a copy of which is also provided on the website.

The first rough list of buildings they planned to connect is there, too. Also, drawings of the three different cable routes needed to get to the buildings, plus some images that explain the advantages of the alternative fiber routes we had in place in downtown Houston. There's an illustration showing how the fiber could be cut and the network would stay up. The concept of Point-of-Presence is detailed in an image we developed in 1986 showing fiber going into a building and to a floor to create a POP to put the equipment in so we could get to customers in the buildings.

There are photos of Scott with City of Houston Mayor Kathy Whitmire and various city council members, including Rodney Ellis (now a State Senator from Houston) and Houston Telecom Director Leonard Childress with Sylvester Turner, a young politician in 1989 who would become the City of Houston Mayor in 2016. They all came to the opening of MFS Houston, an honor not lost on both the customers and political players in Houston. This was the fiber optic network that would launch Houston into what was then called the "Information Age."

Scott also shot numerous videos to document these accomplishments. It is our intent to make these available on the website.

Some early memos concern construction schedules and revenue forecasts and show what the city director dealt with day to day. Kiewit built the network, but Scott, as City Director had to help and make sure they were getting building access rights and other critical minutiae. Building access became a major issue, since there was no revenue generating network if they didn't get into a building or failed to secure the rights to connect the customers or tenants in a building.

Selecting the customers

The job of selecting customers required researching customers in each building, and hiring a sales force to call on them This was an unusual concept because sales must focus on selling only to clients in specific buildings where the MFS presence existed. It was counterintuitive to sales people who naturally wanted to immediately call on the biggest customers in the city. MFS did not have fiber services to sell to Compaq Computers, for instance, because they were not in downtown Houston. It was constantly necessary to remind gung-ho sales reps when they were itching to call on Compaq that it was a waste of time. The fiber was about 20 miles from Compaq's campus.

This difficulty experienced by eager sales people in focusing on serviceable customers seems endemic, as I too have encountered it various times in different companies. When selling any service based on a wired infrastructure, one simply cannot be everywhere. One must sell where the *wires* are. An inability to focus sales on where our wires existed undermined the progress of my later company, NorthPoint Communications Company.

It was a big deal to get fiber up and running. Thus, there was a big turnout for the opening of the MFS Houston Node. The City dignitaries were eager to appear, and corporate people from MFS, including Royce Holland and others who are shown in a photo taken in the node room itself amid an array of working multiplexers. This was December 6, 1989.

In January of 1990, MFS was invited back to City Hall to receive a proclamation glowing with appreciation for the information age infrastructure they had installed. Scott was excited to have the mayor, city council, and key people like Childress acknowledge the importance of the network.

Press takes notice

The media had begun to take notice, too. *Forbes Magazine*[57] published an article explaining the MFS concept in August 1989. This followed on the heels of a May 1989 *Communication Week*[58] article describing how MFS helped a large financial firm connect directly to AT&T without going through the local phone company. This was a startling concept that generated industry shockwaves. It spoke to the change that was happening. A business no longer had to rely on just one monopoly communications provider.

Scott had become part of MFS while the idea of a competitive fiber network was being written about as something brand new to the industry and a big change for businesses. He had been working on the idea since 1986, had won the Houston franchise in 1987, and by 1989 it was no longer new in his mind. The public, however, was just beginning to grasp the concept of competition to the Bell monopoly and how it could help businesses with their long-distance networks and bring down the costs associated with long-distance services.

The articles touched on some of the very concepts Scott had layered into the franchise application with Houston. But not all. He had included using fiber optics for *Local Area Network* (LAN) connectivity across the city, for instance, something no one else was considering. At this moment, it was fresh ground, ripe for upheaval.

---

[57] Forbes August 21, 1989 p.88-89 by Charles Siler "How to bypass your friendly phone company" http://tinyurl.com/RoutingPaths-Forbes89

[58] Communications Week May 29 1989 - Beth Schultz "Bypass Vendor Signs Users" http://tinyurl.com/RoutingPaths-CommWeekMay89

Languishing LANs

Scott approached Mark Gershien, Jerry Brockhurst, together with Kevin O'Hare of Kiewit Construction with the suggestion to create new services. Adding new services to existing fiber was a low-cost way to increase revenue. There were three key items in the fax sent to MFS corporate. This was well before email was a fixture inside companies like MFS. The memo suggested using FiberMUX, which YSA had represented, to do the following:

- Create Ethernet and Token Ring services over the network, on DS3s.
- Use fiber in the risers to deliver services more efficiently and to offer innovative new services.
- Institute a low-cost method of adding buildings to the network, thus making it easier to get more customers to use this new technology in the process.

Thanks to his fiber optic systems integration experience at YSA Enterprises, Scott knew the product, and knew it worked. It was working well for large customers in their campus environments where YSA had interconnected buildings. Scott couldn't understand why the internal MFS audience couldn't see the merit to these new opportunities.

He then realized they didn't understand what LANs were capable of. They didn't understand why customers wanted them. And most of all, they didn't understand the revenue model of LANs. This disconnect between data focused and voice focused engineering is something MFS would encounter many times in the coming years.

Scott found himself failing to convince the internal MFS audience of the merits of this important service offering that generated more revenues over the existing fiber infrastructure for MFS. But the thinking within MFS was mired in the continued ban of electronics in the riser and copper cables pulled to each customer premise. Copper cable is a passive component and does not fail very often or "go down" the way electronics could, MFS managers and engineers reasoned.

They didn't trust the FiberMUX equipment. They didn't believe it was properly engineered in the event a cable was cut or a card failed. He could not shake them of their belief the service would "go down." It was not "carrier class" so the engineering department was opposed to it.

New services were important to the sales force; they could make more money off the fiber that existed in each city. This reflects one key concept that Scott understood very well because of all that "crazy" Lotus 123 spreadsheet modeling at YSA with Larry Koonce. It is very important to create incremental net new revenues for as little incremental capital outlay as possible. Each city director is like a small business person trying to convince the MFS money people called "Development" to commit more money for their city, to expand and land more customers. There was competition for such capital outlays as each city director was competing with the other cities who were doing the same thing.

That there was no way they were going to re-engineer the way services were delivered and they would not inject new technology. *The core business model was based on delivering DS0, DS1 and DS3 circuits which connect the customer to their long-distance carrier and bypass the local phone company. Data was not on the radar!*

At this time, I was at Tymnet and there we had often joked about the distinction between those who approached the world from the perspective of the telephone company verses those who were data centric. We joked about the "Bell-shaped heads" in the Telephone world. The philosophical divide between the two camps was sharp. Data was just a poor stepchild in the telecommunications world, and they believed it would always be so.

We raise the memo not to lament the missed opportunity. MFS was at that time an organization that was "just-like" a telephone company, only smaller and more efficient. They had plenty of people for whom data services were not serious. Scott would probably not have expressed it that way, but he was butting heads with the classic telephone mindset.

The purpose of sharing this today lies in showing that in late 1990 Scott was touting MFS as able to create new services for LAN connectivity that would significantly increase revenues. His fiber optic background served to push MFS engineering and marketing people to explore less expensive ways to add buildings to the network for a very good reason: money.

Doing so would expand the product line and enable new revenue streams. This was an important potential new tool for the sales force. With a broader product line, they could make more sales, and thus more money on the existing fiber in each city. Thanks to the YSA Lotus 123 spreadsheet modeling experience, Scott understood this in a way few others could.

Expanding into LAN services and cheaper building connections would result in winning MFS a greater customer share in markets that were growing more competitive by the week.

Scott learned early on, as this biography shows, the wisdom of listening to customers. Now MFS customers were making it clear they wanted more services than the "Bell-shaped" DS0, DS1 and DS3 circuits.

 Scott wrote another memo[59] in October of 1990 when Jerry Brockhurst asked why Houston had so many unusual requests for services. The answer was that neither Chicago nor New York nor any other major MFS cities, had to compete with Southwestern Bell Telephone Company. Remember that SWBT had moved to block competition by selling "off-tariff" dark fiber to customers rather than DS3 services. The immediate effect was to keep MFS Houston from getting big orders for circuits such as DS3s.

Scott had written to the FCC and complained, but it had not stopped the practice. MFS was reduced to fighting for DS0 and DS1 orders and watching SBTC snap up Enterprise customers hungry to take advantage of cheap ICB services at pricing way under the prevailing tariff.

---

[59] http://tinyurl.com/RoutingPaths-SpecialApps

System Shakedown

On Tuesday, October 17, 1989 at 5:04 p.m., Scott was on a conference call with MFS city directors across the country when the Loma Prieta earthquake rocked the San Francisco Bay Area. The MFS network in San Francisco did not falter; not a single customer lost their communications link. PacBell, the telephone company, on the other hand, went down hard. One of the participants on the conference call was in San Francisco and suddenly dropped off the call.

Meanwhile, Scott was experiencing tremors of a different sort. MFS had a vice president of sales who was very stern and sought to manage through intimidation. That was Tony Pompliano's style, so this VP presumably figured it would work for him as well. In fact, it worked poorly for Tony, and even less well here. It made for a frustrating work environment and an unhappy team.

Everyone was asked in December to make presentations on their sales and revenues. The format, distributed in advance, called for each manager to address gains in their networks. Then they were to put up a slide showing existing revenues on a monthly recurring basis and to show what orders were moving through the pipeline. Next, all orders that were pending, and how much additional monthly recurring business this represented. Lastly, they were to identify the opportunities they anticipated developing over the coming year.

Dave Wheeler joined MFS in Dallas the day before the sales meeting. MFS had no network in Dallas yet, it was the newest city in the MFS fold. He watched in horror as the rest gave their presentations. The mood was dark and somber. People were being threatened with the loss of their jobs if they failed to make their sales goals. They understood these were numbers often set by someone in corporate who evidently thought all cities were rife with sales opportunities identical to Chicago's. They were not, of course. The RBOCs in each region did things differently. For instance, the Houston team had to somehow counter SWBT's predatory dark fiber initiatives that were stealing the DS1 and DS3 sales opportunities.

People were concerned for their jobs. If they did not have revenues per plan, or a pipeline large enough to meet future revenue goals they would be openly attacked before the rest of the group by the VP for sales. It was a vicious environment.

New York City, under Kathy Perone, was doing a great job. She showed awesome numbers even though her network was not complete. Chicago was the number one city in sales. Boston, under Ron Vidal, was coming along too. Minneapolis did its presentation, as did Houston and Los Angeles and San Francisco. The VP for sales unmercifully blasted each in turn.

When the time for the Dallas presentation rolled around, Wheeler stood up and put a blank slide into the overhead projector. Then he took that blank away and put a new blank slide in the overhead. "This is my Dallas network, then another on saying this is my monthly recurring revenue number," he said, swapping that blank for another. "This is our sales year to date for 1990," he said. He had no network, no sales and of course no revenues. However, his gutsy, funny approach broke the ice and generated sales.

By then everybody understood what he was doing. Wheeler had guts, to attempt such a stunt in that atmosphere! After a few moments of stunned silence, they were all laughing their asses off. It was brilliant, a great way to break the ice. The VP for sales and his boys were laughing right along.

Wheeler generated a lot of orders in Dallas and did a great job and will always be remembered for his ballsy, funny performance that day. We had many gutsy people at MFS who got the job done and did not wait to be told what to do. It was a phenomenal environment even with some of the bad management things that happened. Despite some roughness and frustrations, these people would change the world.

Scott continued to try to generate discussion and draw attention to the opportunities he saw. A lot of gutsy people at MFS got the presentations, and Mark Gershien was making the case that MFS needed to create a new image in the marketplace. Of course, a strong feature of the new image involved creating innovative new services such as LAN connectivity. Scott was more convinced than ever that this would differentiate MFS from the competition in a profound way.

Not long after this (in 1991) MFS sold an Ethernet link to The Chicago Natural History Museum. That got the attention of MFS corporate in Oakbrook, Illinois, and the sales force in Chicago liked the product. They had Mark Gershein's ear and he had Royce Holland's and the key development people such as Kevin O'Hara as well as engineering. It felt like a start, at last. Maybe.

DARK FIBER, AGAIN

(1990-1991)

Tony Pompliano was president of MFS when they started a new regulatory initiative. Tony announced a push to gain access to the RBOC CO. It was a significant milestone in the battle against the telephone monopoly.

Royce Holland took over as president later because Tony was not getting the job done. People were working hard. It seemed Tony's brutal and ineffective management style was counterproductive. Yelling and intimidating people for things they had no control over was harmful.

They made their numbers by delivering on more than was expected from local sales. Dealing with the market and business issues proved challenging enough. Dealing with the internal politics and negative attitudes made for greater challenges.

Scott was considering leaving MFS in this time because of the closed-minded attitudes. Scott's focus and interest was still LAN connectivity, but little interest existed in LAN connectivity as a service offering during 1990 and early 1991.

Anti-competitive ILECs

Scott was very frustrated by dark fiber issues across the country. Large DS3 orders, multiple point-to-point T1 orders, and LAN connectivity were being delivered via ILEC provided dark fiber at unreasonably steep-discounted rates. Thus, MFS could not get the big orders in the Houston market. Dark fiber from SWBT continued to keep them away from the big customers.

This was blatant anti-competitive behavior. Scott had filed complaints with the PUC and the FCC, but the PUC did not block SWBT from the behavior. They continued these offers to the big customers using the ICB argument to side-step the tariff and block competition.

T1s alone are NOT the answer.

Only New York City, Washington D.C. and Chicago had the density of T1s to make a city profitable selling circuits without expansion outside of the downtown area. Houston, Dallas, Los Angeles and San Francisco, on the other hand, had to expand to areas bordering the downtown and reach more buildings to generate additional revenue. The need was obvious to the MFS city directors who were driving their sales people to call on every tenant in every one of the first 30 to 40 buildings in the city with fiber. This local loop market was not the same readymade citywide market that existed for long-distance, and had to be built tenant by tenant, building by building.

They were proving they could get 50 percent of the revenue per building for local loop circuits within about two years. It was phenomenal penetration of the 30 to 40 building market, but it was a limited market. It also took time and did not ramp up fast enough to reach break-even cash flow. Upper management did not want to hear this. They were instructed to sell more, so had to stay focused on selling to tenants in the 40-building market.

Tenacity wins the fight

Scott considered leaving MFS, but Ron Vidal and Phil Hamlin convinced him to stay a little longer. Vidal validated his perspective as a city director, agreeing there needed to be more services and more geographic locations for revenues to maximize the opportunities for growth and achieve a break-even cash flow. Hamlin provided the guidance of a seasoned hand.

Scott leveraged his relationship with HL&P to gain access to a 100-mile right of way ring that wound through the major areas of Houston, the first major MFS expansion outside of downtown. He worked hard to make it financially viable.

At the suggestion of Phil Hamlin, Scott called Royce Holland, now CEO and president of MFS. Scott asked him to come down and visit with customers, which he did. When they visited EDS, Bechtel, *Baylor College of Medicine*, Arthur Andersen and others, Royce heard them all say the same thing. MFS should offer native speed (10Mbps Ethernet, 16Mbps Token Ring) LAN services not just DS0, DS1/T1, and DS3 services. The CEO was now hearing directly what Scott was hearing from customers. Hamlin had coached him well, and it worked. Hamlin was more than a little helpful, since he reported to Jim Crowe and Royce Holland directly. Plus, he was a bit of a rogue and liked to back things that were cutting edge.

Scott had previously invited Hamlin to come down and visit the architecture of the network that made use of add/drop multiplexers with digital cross connects. That was in 1987-88. He demonstrated to Hamlin that he was putting a lot of thought into the business, its potential for future growth, and how best to help that growth, even scale it up as sales occurred without having to hire an army of people to manage it.

This was not the direction MFS went when they started in 1989. It was too new and risky, but Hamlin did not mind looking at the new, crazier stuff. He was the CTO of Kiewit/MFS and had deep experience and communications expertise. He solved tricky communication problems when Kiewit built the Alaska pipeline, installing microwave networks, for instance. His support became critical to Scott's push for innovative progress in data services.

After visiting numerous customers over his two-day stay, Scott tells about taking Royce Holland to lunch at Otto's BBQ, considered the best BBQ in Houston by those in the know. Small and low key, it was located on Memorial Drive close to St Thomas High School, the alma mater of both Scott and his father. Royce was the CEO of MFS and had deep telecom experience. It was just the two of them, with Royce's freshly absorbed customer perspective and Scott's data evangelism.

Scott asked who was going to write the plan for these services. Royce suggested that Scott was uniquely equipped to do so. Scott did not have a business degree but had written various business plans by then, and knew what was required. This was an area where he held unique experience and expertise.

He asked Royce what it would need to cover and how it needed to flow. Royce launched into a rapid-fire dissertation, rattling off what was needed to prove that the concept of high-speed LAN services was a viable business. Scott, sitting there with a BBQ stained yellow note pad, was furiously writing down everything Royce said as fast as possible. He barely ate, daunted at the prospect of presenting something so revolutionary for an audience that was sophisticated and knew the business well.

A LAN PLAN FOR MFS

Scott had just committed to writing the business plan, developing financial models, engineering a solution using FiberMux equipment and determining whether it could make money. It was a daunting prospect, one that held immense opportunity, and immense risk. If successful, it could reshape the industry, if a failure it could end his career.

Royce Holland had given the go-ahead to work on the project and make a presentation to the MFS board of directors. A meeting was already scheduled in Houston later that same year. That became his deadline. Scott was now on the hook to convince the board to do something new, to get them to agree that there is a market and that the demand would grow for something that didn't yet exist. He must make a convincing case to the MFS board.

Meanwhile, business goes on. Sales numbers in Houston must still be met, along with growing building access, expanding the network, and so on. Scott's immediate boss, Mark Gershein, a pure numbers guy, insisted Scott had better make his numbers and he would brook no interference from the crazy LAN services project with getting building access, and otherwise expanding the network. He was not on-board with the project, and considered it foolish and wasteful.

Scott had to manage the sales force well enough to sell enough DS0, DS1 and DS3 telecom services to get orders to make their numbers. At the same time, he was focusing on what he envisioned as the chance of a lifetime. It was like getting the franchise. He had to invent the products, the pricing, and the business justification. Now he would also be tasked with convincing a conservative board of directors to devote the money and resources to do something no one had done. It was Scott's idea alone and he could not trade on the reputation of some industry guru. He had to support it with facts and sell it convincingly based on hard data.

Customers had been saying they wanted higher speed services. It was Scott's idea to deliver that service at 10, 16 or 100 Mbps. The MFS board understood that the phone company bandwidth came in 64 kbps chunks (DS0) with 24 of those filling a DS1 (1.5Mbps). Twenty-eight DS1s became a DS3 (45 Mbps). Twenty-eight DS1 pricing was expensive, and the business justification difficult. Multiplexing over fiber was the only technology that matched the native speeds of LANs and did not force them to be carved into 64 kbps chunks.

Royce Holland invented an analogy comparing a DS0 to a garden hose and a LAN to a water main. Scott embellished that concept over time, but Royce Holland came up with it. People could immediately grasp that a water main provides a mightier and faster volume flow than a water hose. Circuits were slow. LANs were fast. The image resonated with customers and MFS management.

In January of 1991 Scott was pushing MFS to create a group in Texas to lobby the PUC. He sent a memo to Royce Holland and others outlining the regulatory battles and the anti-competitive practices of SWBT so they could use the information as needed.

By February he was deep into developing the LAN Services plan. During that conversation over barbecue he pitched Royce Holland on getting customer input to develop the services and define how they would work. He said this would include their suggestions on technology. He granted a budget needed to move forward.

Scott planned to conduct focus groups. He had long been a believer in the value of getting input from real customers. A market study done by a prestigious consulting firm was not enough. This was especially true when developing a new service, one that would use technology not necessarily developed for a carrier service. This service was more in the nature of an Enterprise-wide solution. Input from Enterprise data networking operators and engineers would be invaluable in going about inventing a new service based on existing technology.

To create a version almost immediately, not at some vague future date, they would have to rely on existing technology. Emphasizing *Commercial-off-the-Shelf* (COTS) technology (as opposed to waiting on future developments to create the technology needed for new kinds of services) was a key concept. He had to get management to buy into it. It was risky. The operations and engineering people did not like creating new services, period, let alone new services that used Enterprise class technology rather than the carrier class technology they were accustomed to. He had multiple fights on his hands if he did not win over the operations and engineering people. Enterprise class did not have automatic switching if a card failed and ran on 110 power. Carrier class would switch 1 to 1 or 1 to N depending of if it was a DS0 or DS1 or DS3. Also, they ran on DC power which came from batteries. This enabled the gear to stay operational in the face of power failures, generator switchovers and related interruptions.

Scott conducted focus groups with key people in the Houston Enterprise and university markets. They discussed using fiber optics with LANs and other high-speed applications such as IBM Channel Extension and FDDI, a new standard at that time. They talked about off-the-shelf solutions. For instance, they could install FiberMux Magnums, Time Division Multiplex Ethernet and Token Ring on a dedicated proprietary 100 Mbps fiber backbone. They also discussed how MFS could use this same equipment to TDM the Ethernet or Token Ring segments onto dedicated DS3s. MFS had fiber backbones in all MFS cities and deployed a North American Hierarchy TDM technology that created DS3s on all its fiber backbones across MFS cities. And this meant MFS could use the existing fiber backbones to carry native Ethernet or Token Ring services. This was a big deal since the company had the ability to create DS3s into all their buildings. They could run Ethernet or Token ring over those DS3s and sell the bandwidth in different chunks.

They discussed Fiber Delivered Data Services using the FDDI protocol at some length, since this was the newest technology at the time. The FiberMux Magnum let us do IBM 3290 extension, Dual Wang word processing Extension. They had the ability to deploy IBM Channel Extension using *Network Systems Corporation* (NSC) hardware. Scott called these services collectively "High-Speed Data Services" that included LAN connectivity services.

While he was focused on deploying multiple types of services Scott believed the Ethernet and Token Ring services would be the most popular simply because all Enterprise networks were already deploying LANs in their buildings and campus environments.

He had good turnouts for the focus groups. Exxon, Baylor College of Medicine, Arthur Andersen he would ask the attendees if they wanted high-speed LAN services between buildings, like those possible on a campus where they already owned the right of way. They did. How MFS delivered the service, how much it would cost, and how they approached such quality of service issues as network security were all major issues open to discussion.

The Internet was not yet commercialized. There are no recorded conversations or focus on how to create service offerings that would commercialize it. (Those conversations would emerge soon enough, however, and directly involve MFS in the creation of the Commercial Internet.)

Scott's notes from focus group sessions identify several topics and question concerning the creation of Ethernet (10Mbps), Token Ring (16Mbps) and FDDI (100Mbps) services:

- Creation of Ethernet (10Mbps), Token Ring (16Mbps) and FDDI (100Mbps) services.
- Could we define as a service a TDM channel for an Ethernet that shows up in multiple locations and is a shared Ethernet of 10 Mbps between all the locations? If so, then how many nodes or locations could / should we share among the multiple locations? The FiberMux Magnums limited to a maximum of 8 locations, but that might be overcome if went from a bridged service to a routed service.
- If we were in a multi-protocol environment (meaning that the clients needed to interconnect buildings and carry traffic from IBM, DEC, WANG, Apple, Novell LANs and even voice, or video, between locations) would shifting to a routed Layer 3 service be a problem? Protocols like DECnet, SNA, AppleTalk, and others were not routable and had to be bridged. Even though they explored creating closed user groups of multiple companies via one driving company communicating with all its clients, it became clear that would require a routed service. The alternative would be attempting to bridge together a bunch of LANs operated by multiple companies. That, everyone agreed, would be *really* bad, although the Internet would do more or less just that.
- If they tried to create a commercial way to let anyone route traffic to anyone else (not via the Internet, which was still non-commercial then) that was not easy, and if they tried to create a commercial way they would still face the multi-protocol problem.
- If MFS decided on a bridged service at Layer 2 it was still necessary to figure out what the rules were for building a closed user group with Ethernet or Token Ring. What would the demarcation point be? What would they do in the event of a network issue? (He had yet to figure out what the network management limitations might be at the edge device, the FiberMux Magnum.) How would MFS define this for customers? And could they enforce it using the available management tools of the day?

- The demarcation point was being moved from a circuit and Layer 1 up to Ethernet or Token Ring on top of the fiber and circuits, which was Layer 2. They discussed going up to Layer 3 and offering a routed service. It would require MFS to manage new kinds of services for delivery at Layer 2. The customers were adamant MFS had to deal with the network management issues between the service offering and their backbones. MFS had to manage the service to the edge of the physical layer. Scott had to create a new concept of how to manage the logical Layer 2 demarcation point as well. (To the extent they could define these issues, they were going to set the right expectations. Not only that, they would avoid stumbling into the trap of a pure telephone circuit mindset.)
- A commercial business relationship had to be defined as part of the service that made it clear what MFS was going to deliver to the user at each location in terms of throughput and bandwidth, as well as some definition of quality around bit error rate or loss that was achievable, measured, and repeatable.
- How would they avoid oversubscribing the bandwidth? Bandwidth would be shared across multiple Enterprise customers. Since a DS3 had the ability to host three Ethernets and one Token Ring at 8 Mbps, or four Ethernets, that was easy to track and keep up with. However, if they were going to let the users share segments of Ethernets across the same metro fiber backbone and it was a shared 100 Mbps FDDI ring, how did they ensure there was enough bandwidth per user, per site, and per logical Ethernet backbone while another customer's bandwidth would be shared across multiple Enterprise customers.
- MFS was going to have to watch traffic patterns in aggregate across the backbone and ensure there was not oversubscription by the multiple users on throughput and bandwidth. This was a source of much discussion and it was very helpful to have customers explain what they would, or could, live with in terms of *Quality of Service* (QoS). This also marked a big change within MFS, which previously monitored only whether a circuit was up or down on a piece of the TDM backbone.
- Could MFS think in terms of how much traffic was going across a TDM DS3 or shared FDDI 100 Mbps segment and whether it was getting full or not? This was not a telephony way of thinking but an Enterprise networking way of thinking. This was a cultural issue for the operation people and the sales people to take ownership of as they created the new services. It was clear they had to get a new type of networking group to be part of this new type of service offering if they were going to behave like an Enterprise LAN backbone and not a telephone circuit.
- The definition of network management was going to have to change as things moved up the stack from Layer 1 to Layer 2. MFS was not going to move up to Layer 3 just yet. It was going to be up to the customer to manage at Layer 3; MFS would manage the services at Layer 2.

The sessions were notable in how well everyone collaborated, providing input while respecting the perspectives of others. They were fortunate to get such a talented group of people to help figure this stuff out, and lucky to have Tim Devine of MFS Corporate take part in most of the sessions. His input and help assured Scott that MFS Corporate would buy into the ideas once developed. Dan Strange from MFS Operations and Engineering also participated as part of the corporate overview and was very helpful. Scott professes to be amazed even today that that he was allowed so much freedom and given so much control over the process. It's a tribute to Royce Holland's trust, Crowe and the Board of MFS trusting Royce Holland, and the overall culture of "get things done" that existed at MFS.

After the sessions Scott decided to put out a Request for Proposal (RFP) based on how the focus group wished to see the services work. The end user customers wanted input, but were not interested in being involved in the actual vendor selection or choice of the specific technologies.

When Royce Holland asked Scott to develop the business plan he also said he could hire a consultant to help. So, Scott hired Stan Hanks. They both liked the idea of using the fiber for more than circuits to interconnect customer buildings at native LAN speeds in one metropolitan area.

CLOSED USER GROUP vs. INTERNET APPROACH

Stan Hanks wrote the RFP for the new high-speed data services business with input and guidance from Scott. They defined the services based on the findings of the Houston focus groups with representatives of oil companies, Arthur Andersen, Rice University, and Baylor College of Medicine.

As documented in the videotaped meetings, they decided to go with a closed user group approach rather than with the Internet's newly emerging model of "one connection to the world." Scott and Stan chose this approach after much discussion. Then as now, security defined quality more than global connectivity.

Fast forward to today and security on the Internet is still our No. 1 concern. Identity theft, Ad Fraud, break-ins, ransomware, spear phishing, hacks and more are rampant. And, as I'll endeavor to explain in depth later in these pages, the idea of developing robust closed user groups to solve current universal concerns about Internet privacy and data security is a viable, largely unexplored, solution 25 years later.

Companies in the 1990s wanted, among other things, to securely connect and conduct business at the employee desktop level. Arthur Andersen fit this category. The national accounting firm also wanted to connect to clients using its own applications. The company wanted a closed user group that only it controlled.

These were new and untried ideas at the time. And they excited Scott in the way original ideas always do. He grew enthusiastic as he considered the possibility of a company using MFS services to interconnect their offices and employees via LANs, with a separate service used to interconnect to their clients, and conducting business in a secure manner via a closed user group.

It's important to understand that at the time Scott was developing the high-speed data network, the Internet was solely a research network controlled by the National Science Foundation (NSF).[60] The Foundation created what was then called NSFNet, and which soon became known as the Internet. NSF funded the entire thing, including the transmission backbone, which linked research universities that were interconnected via regional Networks with names like NorthwestNet (Seattle), SURFnet (California), SURAnet (Washington, D.C. to Florida), Sesquinet (Texas), among others. Researchers at the interconnected universities used email to collaborate.

Guy Almes (at Rice University) and Stan Barber (*Baylor College of Medicine*) were also essentially running Sesquinet. They taught Scott about the Internet and its acceptable usage policy restricting non-researchers, including businesses, from using it without permission. The idea of conducting business in any form, much less retail sales, over the Internet was greeted with shock, horror, and hostility.

As these meetings were occurring in 1991, the commercialization of the Internet, to the extent it loomed at all, was off in the distant future somewhere. In fact, most informed people thought that if anything would launch the Internet into the public sphere it would be its sale by government contract through the NSF to a telecom concern, most likely AT&T, MCI or Sprint.

---

[60] https://en.wikipedia.org/wiki/National_Science_Foundation_Network

The mass introduction of email in the early 1990s marked another milestone. Email, along with access to servers with key files, would drive the Internet's expansion.

Of course, email existed long before it emerged as a business essential in the 1990s. Modern email dates to 1972 when Ray Tomlinson picked the "@" symbol to denote sending a message from one computer to another. Long prior to that, email existed on mainframe computers and users could email within the system. It was networking and the addition of the "@" that changed it. Users of the ARPANET used email long before the public had access, but the NSF rules prohibited business so it was not a factor in the business world until the Commercial Internet emerged.

While discussing email and its appearance with the Commercial Internet, I would be remiss if I did not give credit to my own employer in this timeframe for their own early innovation in business email communications. Tymnet, where I worked from 1980 until joining MFS in 1992 offered a good, secure corporate email service.[61] Internet email did not become common until much later, but I was sending and receiving email routinely in 1980, as were a great many Tymnet customers. This early yet popular business-oriented service, Ontyme,[62] is almost forgotten today, even Google finds few references.

Meanwhile, however, Internet users were concerned that IBM might partner with MCI and end up controlling everything. People wanted an open Internet, not one controlled by government or corporation. NSF was going to pull itself out of the backbone at some point, but nobody knew how or when.

Business simply did not take the Internet seriously. It was a playground for academics, scientists, and government, operated for the benefit of few and paid for with tax dollars. There were few ways to make money from it, and the lack of security was considered dangerous.

Scott listened to the discussions but stayed focused on creating a new kind of service for LANs. His experience interconnecting buildings using Ethernet, Token Ring and IBM mainframes to terminals and IBM system 36's made it clear the services needed to be transparent to users. He also knew the FiberMux boxes and other vendors delivered services between buildings in a transparent way so users thought they were on the same LAN inside a building even though they were across town and miles away from each other.

Scott thought of the Internet as a future high-speed data connection for a different kind of user group. In mid-1991 he was focused on how to sell services and products to end users. Customers were looking to connect multiple locations together to use corporate applications. The new type of high-speed connection was far superior to connecting to carriers' low-speed DSx based services.

The Internet was just another high-speed data networking opportunity, a vertical market that Scott thought could be big if it caught on. He wanted to help startups get going but didn't want to be in the Internet business. He thought MFS should remain neutral and provide solutions to all comers.

---

61 http://tinyurl.com/Email-in-ARL-Lib-106
62 http://www.computerhistory.org/collections/catalog/102712811

Thus, MFS decided not to focus on developing an Internet service and to focus on private networking services only. Customers were working with a range of protocols, including DECnet (DEC LANs), SNA (IBM mainframes), IPX (Novell, a LAN server company), and AppleTalk (Apple peer-to-peer computer networking). The TCP/IP protocol existed on for the Internet and was not yet widely used as an Enterprise protocol. Microsoft had NetBEUI, but they did not have networking built into their Operating System (OS) yet. Microsoft was not a real player when it came to networking in 1991-92. It could be purchased, however, at a cost of $150 per computer. Called "Windows for Workgroups," it allowed connections between PCs. I can recall having to purchase the TCP/IP Stack for a PC and spend more than $300 because Microsoft Windows did not have TCP/IP built into it.

While working on the business plan for MFS, Scott also had to meet the sales numbers in Houston. Which meant hiring and managing sales representatives, making sure weekly management reports were completed, and calling on customers as needed to help the sales representatives win orders. He also had to help get building access or MFS could not deliver services. He had an expert on staff focus on building access. Once he made contact and was about to close a deal he'd bring Scott in to meet the building owners or managers. It seemed almost a cultural requirement to meet with the highest level local person before they'd do a deal.

Scott managed the sales representatives and interfaced with the operations people if they were having trouble delivering services. MFS installed a FAX system to process orders. You filled the form out by hand and faxed the form to Illinois where it was compiled and re-typed on an IBM PC that generated the order information. Illinois would shoot reports back to each city. It all seems so quaint today, but if the information didn't get to corporate in Oakbrook nothing would happen. This was a different process than the one at YSA where they dealt with orders for equipment, cables and connectors on the spot.

Scott had to develop his own reporting system to keep sales management abreast of orders and where they were in their quotas as well as manage the sales people so they made their sales numbers each month. If they were not making their numbers, was it because they were not good sales people? Or were they having other problems with for example building access, installation, or sales. Or perhaps a potential customer wanted to stay with SWBT and was hesitant to separate from the powerful monopoly.

He was also busy trying to expand and extend the downtown franchise network to reach 40 more buildings outside downtown. There were many potential customers in Greenway Plaza, the Galleria, the Energy Corridor and the area around the airport, and all were reachable via an 80-mile ring around Houston where Scott had negotiated access to via HL&P rights of way.

LAN's the Plan

It wasn't juggling so many balls that boggled Scott while developing the business plan and pushing this crazy LAN interconnection services idea to its logical conclusion. He has talked at length about the craziness of those days:

[---]

It was the question, "What was I doing?" And it helped to remember how often in my early career I discovered gratification in developing and pushing forward with risky new ideas. At Compressor Engineering Company (CECO) I successfully generated new business by proposing we make parts that we had never built before. And they had to work or we might run the risk of destroying a compressor or killing somebody.

From the compressor industry, I entered another industry about which I knew practically nothing – the wire and cable industry. That lead to copper plenum cable and from there I took the leap into the new world of fiber systems. Fiber systems had to work, too, or I'd be back at square one. There was no faking it when doing fiber systems. Customers would know there was a problem before we did. Thinking up the idea and building a fiber network that could compete with a multi-billion-dollar monopoly, and then winning the Houston franchise – that was huge, certainly. Then getting the fiber network built by Kiewit. And getting customers from scratch, convincing them to do business with a new company, and to use fiber optics in addition, which was a new service, that was a big deal. As was having to develop our own sales approaches, defining our products, and persuade customers to buy services from MFS.

And all of it in hindsight seemed to ineluctably lead here, to this LAN services project. It was intimidating and scary at one level, to be sure. But I again did what I had been doing for years, I just decided I could do it. I followed my gut. And I knew I had to do it well so I could move the idea of LAN connectivity over Metropolitan Fiber forward. If I did that, it would mean we were succeeding and we could create this new kind of service in all the other MFS cities.

I first started talking about this idea with city officials and explaining it to vendors before Kiewit entered the picture. I was explaining it to MFS people in 1989. I'd already been working on it for more than five years by the time I got busy on the high-speed data plan for LANs in 1991. Now it was time to take it to finish line. I would not fail. It was time. Once again, I had been given the ball. I called Phil Hamlin and Ron Vidal.

I thanked Hamlin for suggesting I call Royce Holland and invite him to Houston to meet with my customers. Then I asked if he would look over my shoulder to keep me from doing anything stupid. I called Vidal because, as city manager in Boston, he was dealing with the same issues I faced in Houston. It's always a good idea to have more than one pair of eyes looking at what you're doing. Vidal was a great ear to have, and a good and stable person who helped others around him stay grounded.

I needed the benefit of many perspectives. That included engineering and operations people as well as sales people and others in management. The critical perspective remained that of the customer, but there was no dismissing how very fortunate I was to work with good local engineers and operations people like John, Roland, Rodney and others.

These guys worked weekends installing prototypes of Ethernet and Token Ring Metro Connectivity to the Chronicle Building from our Node. The goal was to deploy a LAN protocol video conference application over the fiber via LAN Multiplexers using PCs as video conferencing interface devices. We were using early versions of video conferencing software on PCs. I wanted to demonstrate an application that might be enabled by high-speed LAN services that would *not* work well on circuits. It worked great on fiber. The Houston Chronicle was impressed. But so were our internal people. As icing on the cake, the operations guys said this stuff was easy to install.

It was the first real demonstration of what we could do. That is, if you ignore the 1990 demonstration of how fast our network switched when a fiber was cut or a card failed in a Time Division Multiplexer (TDM). In that instance I wanted to show early adopters of the MFS network that we were better than SWBT, which was still using copper to deliver T1s. Copper cable was a single point of failure for delivering a T1. The MFS fiber optic circuit product was better than the incumbent phone company from the DS0 to DS1 (T1) level all the way up to the DS3 level. Our service was a diversely routed (two physical routes in the streets) and electronically redundant (two or more electronic cards so if one failed the backup would switch immediately over to the other one in 50 milliseconds) service.

We had a vendor do a demo with us using their video conferencing equipment to illustrate how to do video conferencing over our Houston fiber optic circuit-based network. We used this type of customer equipment because it allowed video conferencing to work over 56 kbps or partial T1 like 256 kbps or full T1 at 1.544 Mbps. All this was called the *North American Digital Hierarchy* (NADH)[63] and it had its own set of protocols. The system is also known as the T-Carrier[64] system.

When we pulled a card out in the Node Room you could sit in the conference room and see it pulled. We would be looking at the video conference person who was in the Node Room with the camera pointing at the person who was going to pull the card in the Node Room. When the person pulled the card, the auto switching in our equipment worked and the video went away for less than 1/10th of a second and re- appeared. We were not only proving the video conferencing system worked using circuits, we were proving it worked over our network and if a cable was cut or a card failed it would only interrupt the circuit based version of the video conferencing service for an extremely short period – one-half of 1/10th second – and then the network would self-heal. This video conferencing link was a great way to demo the network and how it could be used for new, useful applications other than access to the long-distance company.

---

[63] http://www.linktionary.com/n/nadh.html

[64] https://en.wikipedia.org/wiki/T-carrier

We were demonstrating what could be done using LAN and computer based protocols which functioned in completely different ways. Roland Freund and his people (especially John Calhoun) were particularly helpful. They made FiberMux boxes work with almost no support from FiberMux, setting them up to run over a dedicated dual path DS3 that could be provisioned anywhere on our TDM North American Hierarchy Network. Not only in Houston but in any city we were in.

We could put the FiberMux boxes on our existing backbone and install Ethernet, Token Ring, and IBM 3090 RG62 coaxial links, as well as dual Wang or other links, over the same DS3 between two, or three, buildings. We could link up to eight buildings if we built a self-healing FiberMux Ring on top of the DS3s. This was radical stuff. No one had done these things, or even proposed them.

This was also important to demonstrate to Royce Holland and other senior management (including Phil Hamlin, who was looking over my shoulder). This equipment could be installed and managed by our *existing* operation employees. At the time, no one in MFS operations knew anything about this LAN and computer based customer premise equipment. People were that accustomed to thinking about phone lines.

[---]

In 1991 MFS at last began to promote using customer premise equipment to deliver new services. In general, it was frowned upon by the Operations and Engineering organizations, and *Network Operations Control* (NOC) people remained wary of what they considered risks. They did not like vendors who were *Customer Premise Equipment* (CPE) vendors in the transport part of the network. Nor did they like not using carrier class redundant switched DC[65] powered boxes in the network. They reluctantly went along with it and that is another testament to the flexibility and willingness to innovate by the MFS people.

Calhoun and Freund were conducting their low-key FiberMux testing at the city level without the awareness of higher management. The Houston Operations people wanted to understand how difficult it was to utilize FiberMux equipment. Did it work? Could they support it. And they went ahead on their own. It was gutsy and innovative on their part.

It was also a huge success. Those guys were "mission critical" to the entire LAN services development project. They made it work at the Ethernet level first. Soon the Houston Chronicle had video conferencing over Ethernet on a DS3 running from their building back to the node. They were using PCs with LAN protocols to make the video conference work along with a newly developed application for a PC with a microphone and a video camera. They made it possible for people to video conference from anywhere on a private business network. This was in 1992! No one was thinking this way.

Certain applications and hardware had to be built at the ends of a building's links built on top of the LAN Ethernet services. It was increasingly clear that the network alone was not important, rather just an enabler for what was. These mission critical applications on top of a network were important, and only such advanced applications justified the underlying network.

---

[65] https://en.wikipedia.org/wiki/Direct_current

Native LAN service was going to be a huge breakthrough for the nascent industry. Most buildings already had offices and conference rooms wired with Ethernet. They sought to prove that video conferencing would be cheaper and easier to use inside an Enterprise if it ran over a LAN rather than a dedicated telecom system.

The key to this lay, in part, in demonstrating the effectiveness of native LAN speed. LAN speed was a function of the cable already in buildings. This was important to end users (computer users) because they wanted fast response time at the screen. And fast meant seeing data on the screen refreshed in less than one second, preferably less than a half second.

IBM published a whitepaper in 1982, titled *The Economic Value of Rapid Response Time*" by Walter J. Doherty.[66] The central thesis was that the response time at the screen of 400 milliseconds would enhance creativity and productivity whereas more than a half a second (500 milliseconds) limited concentration and reduced productivity. This came to be known as the "Doherty Threshold."

Certain applications, particularly desktop video conferencing, were not being deployed for the lack of LAN interconnection services between buildings. MFS solved this problem with the new *Multi-Megabit Data Services* (MMDS) or Transparent LAN services. The applications easily justified the network. The absence of a cost-effective network, however, could choke off the development and deployment of new applications.

Obviously, applications could decisively drive a network's usage. The connection mattered only in as much as what was being done over the network mattered to a company, meaning that justification was driven by reducing cost or improving productivity.

While testing FiberMux and creating Ethernet over DS3s, his education about the Internet continued. Scott was learning about FDDI, bridges versus routers, and the differences between key vendors. The guys put in 10-hour days developing the RFP and when it was ready, sent it out to TimePlex, LLC, Wellfleet Communications, FiberCom, Cisco and other potential vendors.

Each one had strengths and weaknesses. The sessions with customers lead them to the conclusion they needed to bridge because of non-routable protocols such as DECnet, SNA, AppleTalk, NetBEUI and others that were already installed at the Enterprise level. Routing sounded like a good way to go to create services, but it was something Enterprise did, so MFS decided not to route services. That was a Layer 3 service and MFS had decided to stay at Layer 2, which meant bridging.

Since the Internet was not yet commercial, TCP/IP routing was not much in demand. Cisco, for example, hated bridging, and backed out. They were arrogant about not working with a carrier. They sold to Enterprise, government, "regionals," and the research community. Cisco's focus was "customer premise" not "carrier class" equipment.

---

[66] The Economic Value of Rapid Response Time - November 1982 - Walter J. Doherty, Manager of Systems Performance and Technology Transfer, IBM San Jose, CA. http://jlelliotton.blogspot.com/p/the-economic-value-of-rapid-response.html

Wellfleet Communications[67] did both routing and bridging well. And it didn't care about carrier versus Enterprise. Scott met the founder, in Boston and shared a great lobster dinner. Despite founding a successful company, and amassing immense wealth, he drove a three-year-old Oldsmobile Cutlass. He was down to earth, not arrogant in the least.

He remembered the early development of LANs. He knew the inventor of Ethernet. And he loved the idea that MFS "inventing" Ethernet, Token Ring, and FDDI as a service. He and Scott shared so much early LAN and fiber optic history. His company had a faster bus and faster switching technology than Cisco. But Cisco had come out with the first commercial routers and the early NSF backbone Internet players all used them. Cisco had a brand for Internet stuff and Enterprise multiprotocol routing while Wellfleet was newer and was still earning a brand.

Scott wanted to use Wellfleet Communications, but they didn't have FDDI cards. Even though they seemed better than Cisco in terms of the pure architecture, their backplane architecture did not handle enough switching bandwidth yet to do FDDI, Ethernet, and Token Ring simultaneously.

They had determined they needed to deal with the top management of the vendor company selected. Carrier needs were different than Enterprise customer needs and they needed help from the top to get the right functionality built into the boxes to deliver the robust services they had in mind.

Ultimately, they selected FiberCom. They had off-the-shelf FDDI rings, which allowed Ethernet and Token Ring all in the same box. But they offered something even more important. They were willing to develop a new TCP/IP packet encapsulation method that Stan Hanks had developed with input from Stan Barber and Guy Almes. They worked up and published a document to the IETF Internet working groups.[68] [69] [70] This was a critical piece for sharing bridged protocols across the shared FDDI Media.

It is an understatement to say that this seemed like a lot of new stuff to be trying all a one time.

---

[67] https://en.wikipedia.org/wiki/Wellfleet_Communications
[68] https://tools.ietf.org/html/rfc1701
[69] https://tools.ietf.org/html/rfc1702
[70] https://tools.ietf.org/html/rfc2784

Providing Security for Each Customer

LAN services had to be transparent to customers. We had to let them interconnect their LANs the same way they did in a building. But we also needed shared technology to put multiple customers on the FDDI rings and to develop private FDDI networks on those rings. We had to handle all protocols, especially the non-routable ones, for our large customers. In addition, we had to provide security for each customer via private closed user group backbones over a shared TCP/IP backbone.

Scott recalls these challenges:

[---]

This was a truly radical idea. Stan Hanks was brilliant at working this idea. He "socialized" it, sharing the details with others to vet everything that could go wrong and getting the vendor's best people all the way up the line in the company involved in development for us. We did this while we were placing orders and rolling out the services in Houston in late 1991 and early 1992. It was a wild and crazy time. What else was new?

Stan Hanks and I had debated and discussed what the services should look like and how they would work, given the benefit of our customers' thinking. And it was decided the Internet model was not the best for business-to-business communications services. The Internet was open to anyone who connected to it to communicate with others who were connected. We used the restricted concept of "closed user groups" to create security for the Enterprise customer while still operating over shared media. This was very difficult for Internet Protocol (IP) experts to understand. They were used to the whole goal of the Internet being "any-to-any" connectivity.

This Internet model was counter to the concept of Enterprise networking. In Enterprise networking, everyone was supposed to be able to communicate within one Enterprise, but not be between multiple enterprises. Few people understood the need for a service that could provide multiple closed user groups over a common infrastructure. Nothing like it existed. We were pioneering this concept. Only a carrier would want to put multiple customers on the same fiber infrastructure and offer multiple secure "community of interest" or "closed user group" networks between multiple locations.

An Enterprise would own and operate the private network for themselves and they would not let anyone else in. This was especially true if the highest backbone speed was 100 Mbps and some customers wanted to connect at the 100 Mbps speed. They would see connection speeds throttled back to something like 15-, 30-, or 60 Mbps on a shared but secure backbone.

I had to decide which way MFS should go. Tim Devine out of MFS Corporate was probably the only one besides myself at MFS who understood these issues at that time. He deferred to me. I felt confident we had to carry all the traffic that an Ethernet or Token Ring backbone normally would. And also that the locations that one customer wanted to interconnect should be in one group or community.

Stan Hanks came up with the idea to use encapsulating Ethernet, Token Ring and FDDI frames in FDDI as a shared medium. Initially, we looked at the Network Systems Corporation (NSC) proprietary scheme. We found that Centel Corporation in Tallahassee, Florida had taken this Network Systems Corporation (NSC) approach. Bill Price and Jay Westmark were with this regulated telephone company, but they were not like the "regulated" thinkers I knew. They loved it when I called them and said I was creating this type of service offering for MFS and asked if they would collaborate and share their views on how to do it? I gained invaluable insights from their thinking. It was very encouraging. Their views were similar to ours about the need to interconnect customer sites within a city. I no longer thought I was the only one who thought this way. Here were two guys in Florida with a regulated phone company, Centel Corporation, who were working on the same issues.

They had used a proprietary scheme by Network Systems Corporation (NSC), however, and this concerned us. We did not want to be stuck with a single vendor, and Network Systems Corporation (NSC) was not doing well financially as a box provider. We wanted to use something like IP since it was more of a standard. Also, using IP gave us the added freedom to use it for other purposes.

[---]

Once he decided to encapsulate into IP packets over FDDI as our method of accomplishing the closed user group concept, Stan published an RFC to the Internet community. He hoped to convince Cisco and Wellfleet to build the needed hardware. He explained that there were two opportunities here. First, MFS could use this to deliver services. Second, someday Enterprises would want closed user groups formed on a departmental or working group basis inside the Enterprise. The vendors did not respond. In hindsight, it seems possible that the open everything-to-everything IP visibility mindset had already taken hold in the Internet world.

Further, the concept of using "customer premises equipment" to deliver services proved too much, too new. The vendors simply failed to see themselves as players able to compete with the RBOCs and IXCs. Also, MFS was just too small to mess with in the eyes of vendors like Cisco.

Fibercom, on the other hand, understood immediately and built a prototype for us to prove the concept viable. It took them less than less a month and included shared Ethernet services over FDDI.

They were setting the stage for the industry to create Layer 2 closed user groups. You could route across the wide area network and take advantage of any transport service. They were using FDDI as the service medium, but the concept applied to Ethernet as well. Of course, Stan also pointed this out in the RFC.

Stan was alone in suggesting encapsulation of LAN frames into IP to create closed user groups. And they were also unique in applying it to both metropolitan and national services. MFS chose to stay at Layer 2 to make itself transparent and let the customer remain in control of their networks at Layer 3. This also allowed multiple protocols to operate over the service — even "un-routable" protocols.

The Houston city director job occupied Scott's days. So, it was necessary to write the "Houston High-Speed Data Services" business plan at night and on weekends. During the same time the HL&P right of way agreement came up. He needed this right of way agreement to get MFS to approve the expansion beyond the downtown area.

The High-Speed Data Services plan required a great deal of research. He had no marketing help and yet the industry data had to be developed. And he had to show how the customer premises equipment could be used to provide a carrier service offering. Fortunately, the Centel Corporation effort was so similar it alleviated internal concerns, going a long way to assure MFS that they were not crazy, just innovative. They were using a proprietary scheme, however, while Scott and Stan were taking a more Internet-like approach. Scott wanted the solution open so multiple vendors could build products for it.

This concept of using LAN equipment vendors in lieu of traditional telephony vendors such as Northern Telecom, AT&T/Lucent, or DSC was a major challenge he had to overcome with key people inside MFS. Telephony vendors did not understand LANs. Enterprise-oriented customer premise equipment vendors just didn't understand the issues from a common carrier perspective. They only saw using LAN technology for a single Enterprise. They were focused solely on allowing single Enterprise customers to interconnect all employees of the same Enterprise. They understood LAN connectivity in one Enterprise. They could not envision delivering a service to multiple enterprises over one common infrastructure. They discovered that the shared media concept (using FDDI as a "Band-Aid" technology until a backbone faster than 100 Mbps and more robust architecture could be built) was the service customers and the trade press most talked about.

But the service the early adopter market sought was a dedicated Ethernet service using TDM technology from FiberMux. This taught us a lot about the difference between what customers believe and how they behave. There is a perception of security vs. providing real security for a network. Time division multiplexing was perceived to be secure.

Packet encapsulation was perceived to be less secure. Yet both mixed bits from different customers on the same wire.

The MFS board meeting scheduled for July 1991 drew nearer. They'd received quotations and proposals from all vendors. And they'd already chosen Fibercom to provide the shared medium FDDI TCP/IP encapsulation solution. They picked FiberMux for the TDM metro solution.

All that remained was getting the service approved and selling it to the customer base.

Getting It Done

The creation of Scott's business plan—written with input from Stan—reflects the state of personal computing at the time:

[---]

I worked on the plan in much the same way I worked on previous plans. I used a pen to set down the words to the business plan on a yellow tablet. And I used an IBM PC with DOS OS to run Lotus 123[71] spreadsheet software to develop financial models for the services.

In the past I would then have given what I wrote to my secretary to type up, leaving her to figure out what I was communicating through shoddy penmanship and poor spelling. She used an IBM typewriter because PCs did not yet have inexpensive Adobe typeface printers. (There were Adobe PostScript printers and spectacular fonts in the Adobe Originals series, — the fonts alone could run $275 to $370 but they were for the Apple Macintosh, not the PC platform.)

Apple's computers, still called "Macintosh" then, later shortened to simply Mac, utilized better looking print styles and used Adobe. Apple also had the printers to make the format look good. Apples were years ahead of PCs when it came to inserting graphics. The plan as an end document was going to look much better if prepared on a Mac. We didn't have a Macintosh. However, I did have money to hire the plan typed on a Mac. I would hand write it, give it to my secretary, and we'd get it close and then we'd give it to "the Mac lady" and she would take it the final stages, adding graphics and making it look professional.

I was a shareholder in MFS of Houston, as well as MFS's Houston City Director. I was known to the board. They knew I had the idea to build a fiber network to compete with the phone company even before MFS existed. And that I went out and got a franchise even as I was figuring out how to build and operate it. This placed me in a different category from most people who worked for MFS. The board also knew it was my idea to create the new high-speed data services.

Now the time arrived to pitch them on the concept and prove it could work.

The board of directors traveled to Houston for the meeting, which we held in an MFS office. They mingled with the operations and sales people before and after. I drew Ethernet, Token Ring and FDDI diagrams on the same whiteboard we used months earlier to work out the ideas for the services with our focus groups. The board members listened as I unpacked the key concepts. Royce, who had well prepared me for the sorts of things I'd be asked, also stepped in to help with the answers to their good, smart questions.

It turned out to be fun and exciting. They listened, asked questions and were open minded. It was extremely rewarding after years of trying to get these ideas in front of people who could enable them to happen. Knowing what I know today about innovation I am impressed and grateful to have such fantastic leadership.

The board decided to fund the idea.

---

[71] https://en.wikipedia.org/wiki/Lotus_1-2-3

They were responsive especially to my message that, although new, these services leveraged the existing fiber infrastructure. Incremental costs associated with delivering the new services were therefore tied to revenues. Essentially, this meant that MFS need not spend a lot of money up front. We wouldn't put a lot of equipment in a city network unless it was tied to an order that immediately generated monthly recurring revenues. Royce made sure I appreciated how critical this concept was, and why. This was how MFS could keep funding flowing from Kiewit. We had no problem linking revenue to cost. It was part of the architecture of the plan. Spending capital dollars to get monthly recurring revenues that paid off the capital in 12 months or less was a good investment.

It was an entrenched theme, re-enforced by the financial people at MFS who instructed all MFS city directors to focus on getting orders in our existing "on-net" buildings. When we got an order in an "on-net" building we would be spending money to turn up revenues in buildings we already had fiber in. That was a lot cheaper than constructing a fiber network to a new office building. The High-Speed Data Service offering I successfully pitched called for spending money only on those boxes needed to deliver the High-Speed Data Service itself.

Royce and other MFS management drilled into me that the incremental capital cost of turning up new revenue producing orders would be treated differently on our books than capital spent building basic fiber infrastructure in a city. The initial fiber backbone included a major Downtown Node and Points-of-Presence (PoP) in 30 to 40 buildings. The cost to get in the local market was typically $5-$10 million dollars per city. Forty or so buildings constituted the minimum size network needed to profitably sell services in a new city.

Since by mid-1991 MFS was in 11 cities and aiming to add 20 more, it made sense to figure out ways to create new revenue using the same basic fiber infrastructure in each city. This revenue and cost equation was where the board riveted its attention. Customers wanting something new and asking ourselves how we might meet those needs was secondary. But that's okay because they got the point about the customers. And they approved my plan. And they'd agreed to let me keep Stan Hanks as a consultant. Hanks provided an invaluable interface with vendors and customers.

We were using the TDM DS3 approach and the FiberMux product, which put us in a position to deploy Ethernet or Token Ring services. We could deliver as many as two Ethernet and one Token Ring customer links over a single DS3 point-to-point or point-to-multipoint in MFS cities. We also had the approval to buy FiberCom gear when the encapsulation software was completed and offer that as a service as well.

We were assiduously avoiding direct competition with such Interexchange carriers as MCI, Sprint or WilTel because they were our customers for MFS local loop DS0, DS1 and DS3 circuits. Consequently, at startup in 1991 and early 1992 we restricted the sale of our new services to inside the metro areas of MFS cities. We did not then plan to sell them across country. That would come later.

The progressive ideas we were pursuing benefitted greatly from the unstinting support of Crowe and Royce Holland who early on fueled the project with their interest. I'm grateful for their willingness to commit money and resources so aggressively, especially since many experts were predicting everything would stay voice driven and the future was therefore in voice minutes, dial tone, trunks, and local loop circuits.

Turning once more to the approval of the plan, after the board's blessing, it was, as they say, show time. We were off to New York City to announce the new services to the marketplace. We believed a huge number of customers were waiting for just what we now had to offer them in Manhattan and Chicago.

Royce Holland made the announcement at an MFS presentation. I was there, of course. So was Stan Hanks. And so was Stan Barber from Baylor College of Medicine. Houston would be the first city to offer the services.

Ever the gentleman, Royce Holland took time that day to introduce and thank me for doing the hard work within MFS to create and bring the new services to the market. It was a magnanimous gesture and greatly appreciated.[72]

As we described the services, we again talked about them like circuits. We also referred to the concept of "point-to-multipoint," a telephony way of thinking about LAN-like connectivity.

We wasted no time setting to work pursuing customers in Houston for these new high-speed services. These would be provisioned over a shared FDDI backbone that used TCP/IP tunneling, or alternatively used the FiberMux TDM solution on top of DS3s.

Once this announcement occurred, I was shifted from being the MFS Houston City Director to being the National Strategic Sales Director for Data Services. I had no staff. I was a sort of national resource to help the existing sales departments in each city book orders for the services. I also coordinated engineering and operations efforts in each city as picked up new customers.

MFS was adding new cities and expanding its existing networks in others. At the same time, we were moving ahead selling DS0, DS1, and DS3 circuits. This illustrates the kinds of media coverage that MFS was getting in 1992.[73]

[---]

And as that transformation was developing, there were fateful things developing with an Internet-focused Enterprise known as UUNET, and a large telecom company called WorldCom. The first major competitive local exchange carrier was just a few years away from merging with what was a major part of the Internet backbone and the fastest growing ISP in the 1990s, and then merging again with what was for a time the second largest long-distance company in the United States. I would soon join MFS and become immersed in the new world of LAN-based telecommunications. It was an interesting time.

---

[72] Video clip of Royce thanking Scott for making the new services happen
https://youtu.be/Iu9I6r6eHRQ
[73] https://en.wikipedia.org/wiki/Metropolitan_Fiber_Systems

PREHISTORY and the CREATION OF MFS Datanet

Fully comprehending the formation of MFS Datanet requires understanding the history of data communications before the 1990s, and before the Internet. It is important to understand that the Internet did not magically appear overnight. Many significant precursors occurred in the 1960s and 1970s, and a few well before. The historical roots of the modern Internet are wide and deep.

Smoke Signals to Flags to Telegraph

Humans have been communicating over great distances in various ways since ancient times. In his 1998 book, "*The Victorian Internet*," author Tony Standage[74] addresses this topic in a fascinating and entertaining way, from early signaling towers using mirrors and semaphores, through the Morse Telegraph (1838).[75] During the Civil War, flags and lanterns provided local battlefield communications. Abraham Lincoln used the telegraph to good effect during the Civil War, directing his generals from a distance. Telegraph offices became fixtures in the 19th century and early 20th century towns all over the world. Telegraphy became an effective means of long-distance communications, but it had one great weakness. A skilled operator was required to decipher '*Morse Code*' and hand-write the text on a telegraph form to be delivered to the recipient. This skill was rare and there was an insatiable demand for telegraphers to send and receive the messages. Further, no matter how skilled, humans grow tired and make errors. Telegraphy is powerful and effective, but limited by the weakest element, the human being operating the key.

Limited the technology may be, yet it still has the power to surprise. There have been many demonstrations of "antiquated" Morse code competing against modern texting.[76] Millennials might be surprised at how fast and efficient 19th-century "texting" could be.

Teletype, the first computer terminals

In 1902, an electrical engineer named Frank Pearne envisioned a system of remote printing over wires much as with the telegraph, but without a skilled human in the chain. His idea was an extension of the Stock Ticker, invented in 1869. Pearne approached Joy Morton, head of the *Morton Salt Company*, seeking money to develop the idea. Morton, attempting to determine whether the idea was practical, consulted a mechanical engineer named Charles Krum. Krum worked for Morton's brother, Mark Morton who ran *Western Cold Storage Company*. Pearne left the project but Krum continued the work with his son Howard Krum. In 1903 a patent was filed for a "Typebar Page Printer."[77]

---

[74] https://www.amazon.com/Victorian-Internet-Remarkable-Nineteenth-line/dp/162040592X

[75] https://en.wikipedia.org/wiki/Electrical_telegraph - Morse_telegraphs

[76] One of many YouTube videos demonstrating the effectiveness of Morse Code. https://www.youtube.com/watch?v=64tfnG77Nl8

[77] "*U.S. Patent 888,335 issued in May, 1908*"

In 1906 the *Mokrum Company* was formed and in 1910 the first commercial Teletype system went into service between Boston and New York City. In 1928 the company became Teletype Corporation and in 1930 the company was purchased by AT&T.

During the 1930s and 1940s the Teletype came into prominence, and it became possible to send "wires" "cables" and "telegrams" all over the world. These networks had one thing in common that is very different than today's Internet, besides the speed and technology; someone owned them and controlled them. Marconi, Western Union, various governments around the world, everywhere they existed, someone was trying to control the communications networks of the day, and long-distance communications were limited to the few.

That is a tremendously important point.

Another key point we often lose sight of, is that network communications were in plain text and visible to any number of parties along the way. Sending secrets by these technologies was impossible. This was the accepted norm and no one building early computer networks considered it important to do differently. Various ciphers and codes evolved because of this.

Computers and Telecommunications

By the 1960s, computer manufacturers began experimenting with telecommunications. Teletype machines, in various forms, became the primary mechanism of remote computer access. Unix,[78] for example, was specifically written to use a Teletype. In the late 1960s and early 1970s it was common to spot a computer tech lugging around a sixty-pound model 33 ASR[79] complete with paper tape. Many backs were spared when Texas Instruments introduced their ubiquitous Silent 700 terminals complete with a built-in acoustic coupler in 1972 and for nearly two decades they were commonplace fixtures among data communications "Road Warriors."

In 1964, IBM introduced the 2848 *Display Control Unit*[80] that could connect to up to 24 IBM 2260 Display Stations. The model 2848 could optionally interface to the Western Electric 202d (1200 bps) or 201b (2400 bps) data phone,[81] thus connect the Display Station to the Display Control Unit via telephone lines. The 2848 gave way in 1971 to the more powerful and more flexible 3270[82] and by then many companies were connecting remote terminals to their computers. The 3270 systems yielded to IBM's *Systems Networking Architecture* (SNA),[83] introduced in 1974. The 2848 and 3270 systems connected remote users, but before SNA there was no attempt to connect the computers together that they may communicate with other computers. The SNA extension APPN[84] added this capability, although that came later.

---

[78] http://www.linusakesson.net/programming/tty/index.php
[79] https://en.wikipedia.org/wiki/Teletype_Model_33
[80] http://ed-thelen.org/comp-hist/IBM-ProdAnn/2848.pdf
[81] http://www.smecc.org/modems_and_acoustic_couplers.htm
[82] https://en.wikipedia.org/wiki/IBM_3270
[83] https://en.wikipedia.org/wiki/IBM_Systems_Network_Architecture
[84] https://en.wikipedia.org/wiki/IBM_Advanced_Peer-to-Peer_Networking

The use of 2848 / 2260 with Western Electric data sets constituted the first computer networks, of a sort. The technology was primitive and limited to well-heeled businesses. Nonetheless, it constitutes an important milestone in networking history rivaling the introduction of the telegraph and the Teletype in importance.

The IBM 2848 system contributed another technological innovation to the data communications industry. As the first system to communicate over a phone line, it was the first system to encapsulate its data into a primitive form of what were later called "packets" and this "2848 protocol" long outlived the 2848 system itself. The 2848 communications protocol found a home in two major markets, airline reservations, and credit reporting because these industries established themselves on early IBM systems. The *Semi-Automated Business Research Environment* (SABRE)[85] grew out of the US Air Force *Semi-Automatic Ground Environment* (SAGE)[86] project.

American Airlines started the project in 1953 when they first considered a proposal by IBM. In 1957, after four years of research into the practicality, a formal development agreement was signed and the first experimental system went online in 1960. A similar series of events occurred in the credit reporting industry when IBM pitched a similar idea there.

Packet Switching

 The next milestone came in the 1970s when *packet switching* appeared. Paul Baran of Rand Corporation published the seminal research[87] between 1960 and 1962. The history of ARPANET has been written many times, but another early networking story has not been told nearly as often; that of the first *commercial packet switched networks*.

The 1960s computers spawned many businesses intending to leverage the growing power of the new machines. IBM was the dominant computer manufacturer, but there were many other players. In 1961 *Scientific Data Systems* (SDS) was founded to build a computer using then brand-new silicon transistors. In 1966 they shipped their newest, most powerful system, the SDS 940.

Meanwhile, Tymshare, Inc. was founded in 1964 to offer time-sharing of computer services. Their early attempts to connect remote users used acoustic couplers designed by Anderson/Jacobson[88] expressly for Tymshare. This had the drawback of being slow (110 bps or 300 bps, depending on the modem), unreliable and expensive.[89] Long-distance calls were expensive and the slow data rate meant sending data was expensive. In 1966 Tymshare acquired an SDS 940. In 1968, Norm Hardy[90] and LaRoy Tymes[91] conceived the idea of combining telephone lines and the communications power of the SDS 940. That idea spawned a networking technology

[85] https://en.wikipedia.org/wiki/Sabre_(computer_system)
[86] https://en.wikipedia.org/wiki/Semi-Automatic_Ground_Environment
[87] http://www.rand.org/about/history/baran-list.html
[88] http://corphist.computerhistory.org/corphist/documents/doc-419e6fc2c2c69.pdf
[89] https://en.wikipedia.org/wiki/Modem
[90] http://www.computer-history.info/Page1.dir/pages/Hardy.html
[91] http://www.computer-history.info/Page1.dir/pages/Tymes.html

known as Tymnet. There were several experimental designs and various attempts to build a network and by 1971 they had the network operational.

The first full-fledged version of the system came online in November of 1971 serving Tymshare customers. Three months later, February 1972, Tymnet began carrying traffic for the *National Library of Medicine*. This became the first external, non-Tymshare, paying customer to buy independent data transport service on the network and marked the very beginning of the commercial data network. In 1979, Tymnet, Inc. was "spun-off" from Tymshare and became a public Common Carrier.

Meanwhile, in the ARPANET realm, *Bolt Beranek and Newman* (BBN) had built much of the hardware and software for ARPANET. When Tymnet went operational they decided they had the technology and resources to do the same. A commercial Data network known as *Telenet Communications Corporation* (Telenet) came online in 1974[92] and began offering commercial services on August 16, 1975, to paying customers.

This late 1970s era was a period of innovation in packet switching that extended into the amateur realm. Amateur Radio operators began experimenting with packet-switching on the airwaves, creating something called AMPRNet[93] in 1978. I was an active ham at the time and had my own AMPRNet station on the air.

Telenet is often credited with offering the first commercial packet-switched network. That claim is questionable, as Tymnet beat them by more than two years. Telenet was, however, the first using ARPANET technologies, and TCP/IP specifically. In the narrow view of many, this is the only true packet switching.

Tymnet used a similar, although proprietary packet switched protocol that was not publicly defined. Despite the opinions of some, it certainly was packet-switching and was in many ways quite similar to TCP/IP. Discounting Tymnet's networking success does a great disservice to many good people. There is no question in my mind that Tymnet was the first true commercial packet-switched network, by a wide margin, and that Tymnet stayed technologically ahead of the competition for nearly two decades. Tymnet possessed unrivaled translation and conversion capabilities and had a business-focused, world-wide footprint that no one else could touch. Unfortunately, the advent of commodity Commercial Internet services from the ISPs connected via MFS High-speed LAN Interconnect services undermined the business-focused model of Tymnet.

For several years Tymnet and Telenet competed in the marketplace and became widely known as the "T-Net Twins" from an article in Forbes magazine.

Tymnet and Telenet were very different from today's Internet. Not just that speeds were slower and the technology seemingly primitive to modern eyes. Tymnet did not support the unrestricted "any to any" communications. They were strictly business, strictly commercial traffic, and strictly subscriber to provider. Credentials of a sort were required to log in to the provider. Modern X.509 credentials did not exist of course, but you did need a username/password combination to connect to your destination, in addition to whatever login was required at the host service provider.

---

[92] https://en.wikipedia.org/wiki/Telenet
[93] https://en.wikipedia.org/wiki/AMPRNet

Traffic was not encrypted, but nonetheless the service was quite private. There were no points at which the data could easily be intercepted, and there was relatively scant opportunity for bad actors to invade.

Most services were subscription only, e.g. Two significant examples of this were the *Bank of America,* which offered *"Video Home Banking"* via Tymnet in the early 1980s and *TRW Information Services,* which used Tymnet to distribute its credit database information to subscribers. TRW used the IBM 2848 protocol, and Tymnet developed the concept of "tunnels," much like modern VPN tunnels, to transport the IBM protocols over the packet switched network.

Tymnet also incorporated within the network a sophisticated "Protocol Conversion" capability that allowed a simple teletypewriter to connect to an IBM 2848 Display Control Unit where the network would perform the necessary translation so that the IBM "saw" a 2260 terminal. As the network matured, the matrix of protocol conversion capabilities became stunning.

Both services developed into large private networks (BOFANET and TRWNET) built on Tymnet technology. I joined Tymnet on February 15, 1980, and soon became primary tech support on both networks, as well as participating in the sale, installation and ongoing support of several more. Networks installed during this time included Uccel Corp, Dialog Information Services, Southwest Bell, and Chilton Credit (absorbed into Experian). Tymnet grew the data communications business aggressively throughout the 1980s and the private Tymnet-style networks were a major contributor to the bottom line.

It became commonplace to "log in" and retrieve information with ease using a portable Terminal such as the Silent 700. *TRW Information Services* later became credit giant Experian. A bank customer with a terminal or a computer and modem could log in and do their banking online much the same way they do today via the Internet. Tymnet offered an extensive suite of services including email, transaction processing, file transfer, protocol conversion and more, long before such services were commonplace on the Internet. Tymnet customers offered their subscribers an amazing array of services too. Many of the services we associate with the Web today began in this world of Teletype-driven online networks, with roots in the 1903 invention of the Teletype machine.

Early computer networks had their hackers and security concerns too, and throughout the 1980s customers of these services were often the target of hackers and espionage. A particularly noteworthy 1986 incident involved the KGB, an astronomer at Lawrence Berkeley National Laboratory, and numerous Tymnet folks, including yours truly, Nathan Gregory.

An excellent book by Clifford Stoll, *The Cuckoo's Egg* (1989)[94] provides an entertaining look at the issues of online security facing 1980s network users in what is called "a computer-age detective story." The book was also turned into a movie in 1990, *"The KGB, the computer and Me."*[95][96] Highly recommended.

---

[94] https://www.amazon.com/dp/B0083DJXCM/ref=dp-kindle-redirect?_encoding=UTF8&btkr=1

[95] http://www.imdb.com/title/tt0308449/

[96] https://www.youtube.com/watch?v=EcKxaq1FTac

Open Systems Interconnection

Another network architecture that emerged in the late 1970s that would have created a comprehensive set of standards for computer networks called *Open Systems Interconnection* (OSI). Its architects were a dedicated group of computer industry representatives in the United Kingdom, France, and the United States who envisioned a complete, open, and multilayered system that would allow users all over the world to exchange data easily and thereby unleash new possibilities for collaboration and commerce.

For a time, their vision seemed like the right one. Thousands of engineers and policymakers around the world became involved in the effort to establish OSI standards. They soon had the support of everyone who mattered: computer companies, telephone companies, regulators, national governments, international standards-setting agencies, academic researchers, even the U.S. Department of Defense. By the mid-1980s the worldwide adoption of OSI seemed certain.

However, OSI, like the prior closed networking models, stalled in the face of a truly open alternative. Despite its title, the OSI systems were not very open. They were tightly controlled under the umbrella of the International Standards Organization (ISO). Those seeking to build systems using OSI protocols were required to pay licensing fees. TCP/IP was openly published, freely available and could be used by anyone. As OSI faltered, one of the Internet's chief advocates, Einar Stefferud, gleefully pronounced: "OSI is a beautiful dream, and TCP/IP is living it!" The influence of the OSI effort still lives in the OSI "Layer Cake" that maps the Internet Protocol[97] stack into the OSI model, though an imperfect fit.

CCITT X.25

X.25 is an international standard protocol suite for packet switching. X.25 is often thought of as an OSI network protocol, but in fact it is not. It emerged in the 1970s drafts from the CCITT (now the ITU-T) that became finalized as *The Orange Book* in 1976. Unlike OSI, it was conceptualized as a three-layer model. X.25 became popular in the 1980s and for a time was widely supported. Tymnet offered extensive protocol conversion capabilities to X.25 customers.

Today many erroneously believe that Tymnet was an X.25 based network. It was not, this is a misunderstanding. Tymnet supported X.25 thoroughly. I moved to San Jose in 1983 partly because of Tymnet's strong support of what seemed then to be an important service, as I saw value in becoming an expert in the field and wanted to be near the action. In the coming years, I would install many X.25 interfaces. Tymnet supported X.25 and provided protocol conversion services between X.25 and many other protocols, but did not use X.25 internally.

Today X.25 is essentially replaced by TCP/IP, although as late as 2012 it still survived in the credit card payment industry. In the early days of MFS Datanet, some consideration was given to building the network on X.25. Fortunately, more reasonable thinking prevailed.

---

[97] https://en.wikipedia.org/wiki/OSI_model

British Telecom and PKS spawn MFS Datanet

By 1990, Tymnet and Telenet were legacy networks, burdened by obsolete hardware technology and a huge customer base resistant to change. One reason for that reluctance was the customers' reliance on Tymnet's unrivaled protocol conversion capability. They could connect virtually any terminal or device to virtually any host provider, handling all conversions in the network. It was a potent capability.

Tymnet was sold to British Telecom, who sold it to MCI, which was then sold to WorldCom. WorldCom also later bought MFS. Telenet was sold to GTE. The legacy business model was thriving still, but change was in the air. The Internet was on the horizon and some thought it would soon obsolete the legacy networks.

We lose sight of the simple reality that this was not nearly as obvious then as it seems in hindsight today. Many believed in the Internet, but the financial picture was unclear. Networking was of interest to business and business had the money to pay for it. The Internet prohibited business and commercial activity. Without business, the Internet did not have a sustainable business model and depended on government largess. The viability of the Internet as a business model for a networking provider was questionable. Many, including many of the Tymnet folks who later joined MFS Datanet, believed it was folly to court the Internet Community as customers. Little money and an unsustainable business model raised red flags. Then. Few saw the coming changes on the horizon. The juggernaut that it would become was not evident then.

Tymnet missed the significance of the Internet almost entirely. Toward the end, some folks within Tymnet were playing around with a cobbled-together Internet Gateway maintained by Steve Feldman and Paul Krumviede. Usenet saw a lot of internal use thanks to their efforts. It was very much a skunk-works sort of thing, as no one was taking the Internet very seriously.

This was pre-web, before even the Mosaic browser existed. The tools of access were not browsers but were FTP and Telnet, and Gopher, Archie, Veronica and Jughead were search engines. Destinations included DECUS, and instead of Facebook, we had Usenet and email. It was all text-based and slow, but there was even then a great wealth of information to be found if you knew where to look.

Wayne Flagg was spending his time creating Unix-based tools to manage the Tymnet network. Both Steve Feldman and Wayne Flagg soon joined MFS Datanet. Paul stayed with Tymnet after the MCI acquisition, ultimately working for Vint Cerf, where he made numerous contributions to the development of the Internet. His name appears on RFC2095,[98] for example. We who knew and worked with him were saddened to learn of his untimely death. Paul died March 26, 2004, at age 42.[99]

---

[98] https://tools.ietf.org/html/rfc2095

[99] http://www.legacy.com/obituaries/name/paul-krumviede-obituary?pid=1000000002081372

When BT bought Tymnet, the President of Tymnet was Al Fenn. Al had been with the company since the early days and held many positions including VP of Development and General Manager. Al Fenn was given the title VP of Mergers and Acquisition for *British Telecom North America* (BTNA) and tasked with finding suitable acquisitions for BTNA in the states. One of the first potential acquisitions he pursued was a certain tiny telecom company owned by a huge construction company, *Metropolitan Fiber Services* (MFS).

Al tried several times to coordinate a meeting between the PKS board and the BT Executives. The BT people were reluctant to pursue the opportunity, ultimately the Kiewit people offered Al a job. They had been experimenting with data networking under the umbrella of MFS Telecom, but telecom engineers did not understand data networking. Kiewit wished to create a new company under the MFS umbrella, one that understood data networking and could pursue it independently of the Telecom.

Al worked for PKS as a consultant for a time, before deciding to leave his position with BTNA and come on board full-time. MFS Datanet was born to replicate the formula of Tymnet's business-oriented data communications service on modern *Asynchronous Transfer Mode* (ATM) technology, building a modern, much faster version of Tymnet.

Al Fenn left BT Tymnet to found MFS Datanet, ultimately taking more than 30 employees with him, including me, plus Ken Holcomb, Ron Whitlock, Steve Feldman, Bill Euske, Dan Lasater, and BJ Chang. That was just the first wave, and many more would soon follow — so many in fact, that some feared a lawsuit from BTNA over poaching of talent. That did not happen, and in fact, BTNA thanked Al for helping them reduce headcount as Tymnet was being prepped for the sale to MCI. No one cared that much of their important brain-trust was walking out the door. Meeting headcount goals was what mattered.

Datanet provided the missing piece MFS needed to blend data services into their service offerings. That which MFS Telecom engineers resisted, MFS Datanet eagerly embraced.

Developing MFS Datanet

Scott spent the rest of 1991 and early 1992 visiting customers in Houston, but also traveling to Chicago, New York, and D.C. to visit Enterprise or government customers. He was evangelizing the High-Speed Data products and services of MFS. Stan Hanks accompanied him when he visited the key Enterprise customers in New York. They were the most sophisticated customers and recognized response time translated to money. They understood that having a 10 Mbps Ethernet link was much better than a T1 and was much cheaper than a DS3. Not only was a DS3 more expensive, it was inefficient. It wasted most of the 45 Mbps bandwidth when dedicated to a single Ethernet connection. The MFS Fiber solutions were much more efficient and more cost-effective.

The response time for their specialized analytical applications had to be fast. They wanted their employees to make key trading decisions and execute on them faster than their competition. Native LAN speed connections allowed sub-second response times, far faster than competing technologies.

A key feature of the service was that it was not shared. Each user had access to the full bandwidth of the interface and could use 100% of that bandwidth without impacting anyone else. This turned out to be important to a new class of customers we were then courting, *Internet Service Providers* (ISP). They saw our service as backbone trunks upon which they could transport data streams from many individual subscribers. One of those first ISPs approached was Rick Adams of UUNET, which had been founded in 1987 and then was one of the first Tier 1 networks.

In his year-end report of December 1991, Scott outlined to Royce how Rick Adams wanted to be "on net" in order to buy the MFS 10 Mbps HLI service.

 A memo[100] by Dan Strange of January 20, 1992 outlined what they needed. He proposed creating the first Metropolitan Area Ethernet service to interconnect Private IP Backbones. These are the companies who became known as *Internet Service Providers*.

The January 20th memo listed the pipeline of deals with item 9 being UUNET, the service that later became MAE-East. Scott was VP of Sales and was listing all the metro point-to-point or point-to-multi-point LAN backbones that were Ethernet, Token Ring, and FDDI deals in the pipeline. While working on metro data deals, they had to build the national backbone, since MFS planned to sell High-Speed LAN Interconnect (HLI) services between cities, not just in one city.

PKS and British Telecom

During the second half of 1991, Jim Crowe decided there should be a new business unit based on the High-Speed Data Services. Jim Crowe began searching for the right person to head up such a business unit.

---

[100] http://tinyurl.com/RoutingPaths-Memo-MAE

Fortune decreed that the idea for a new data-centric business unit intersected with the PKS talks with *British Telecom North America* (BTNA). As explained in the previous chapter, PKS was exploring the sale of MFS to BTNA, but that plan failed. They recognized that BTNA had a network much like what Scott had in mind, but Tymnet was hardware moribund, slow, and BTNA lacked the vision. The serendipity was too strange and strong to ignore.

Al Fenn understood business-focused data networking. PKS hired AL and much of his team. They brought Al aboard, first as a consultant to develop a plan and when that became solidified, Scott was tentatively offered a position working for him as the VP of Sales, though he had to sell Al on his ability.

At one level, this was a frustrating case of *déjà vu*. When Scott championed the idea of building a fiber optic network in Houston, he landed the deal; MFS bought it and offered him a position under Tony Pompliano, for which he had to interview. Again, Scott had done all the hard work on the High-Speed LAN Services and MFS was bringing in someone he must sell himself to.

Scott met with Al Fenn, his right-hand man Ken Holcomb, and Stan Hanks in early 1992 just as they were forming MFS Datanet. They discussed the need for a national data service. Everyone felt strongly that some company should develop such a national service. Scott explained to Al the reasons why he would find huge resistance to the concept within the ranks of MFS. It sold to IXCs which would consider it competition if MFS were to sell a service that eliminated the need for buying a long-haul circuit.

Scott had been working with Sprint, MCI and ANS[101] to figure out a way, using the encapsulated IP approach, to have MFS be the intelligence on the ends of national DS3s. He did not think it possible to get the ability to offer national services approved within MFS until much later. The opposition he saw was great.

Al floored Scott when he said he already had approval from Jim Crowe and the board to offer national data services. He also said he had decided to investigate using *Asynchronous Transfer Mode* (ATM) technology. It seemed a natural choice to deploy these services. He wanted to do further research to make certain the technology was sufficiently mature. MFS had been the first carrier to offer a high-speed data service in the Houston metro area that used TCP/IP over FDDI as the network layer protocol. They had used the FiberCom platform with the IP encapsulation developed by Stan Hanks. Now Al was going to change and turn away from that approach.

Scott felt impressed Al had pulled this off within MFS. He thought the MFS insiders would not have listened to him on the issue of a national backbone service offering. Scott feared to even try because it seemed like such a big decision and fight within MFS to take on. Scott was amazed that Al could get the ear of key MFS people. What Scott didn't realize was that PKS had recruited Al expressly for that task after the failed sale of MFS to BTNA. Al had successfully pitched the PKS board on the idea of reinventing Tymnet using modern technology, and once convinced, they had given him carte blanche. In any case, Scott was elated at being allowed to develop and deploy a national high-speed data network.

---

[101] https://en.wikipedia.org/wiki/National_Science_Foundation_Network

Still, he had been reminded that he was an employee of MFS who happened to own some of the stock of the company. He was the founder of NCI but not of MFS. He had originated the idea to create high-speed data services but they had brought Al Fenn in to develop it as a business without his counsel.

Scott focused on the exciting opportunity before him. He reminded himself how fortunate he was to be able to build the network in Houston because of Kiewit and MFS. He reminded himself how great it was to create businesses that no one had ever built before and have the money and resources to do it.

He had been almost destitute enough times in his own business and while developing the metro fiber optic business concepts, that he did not take it for granted being part of this great adventure. Al Fenn was clearly more experienced than Scott and with strong people ready to build the organization that was essential to the success of the business.

Al, Royce and Jim Crowe kept Scott working on the high-speed data services once Al was brought in to run it. Jim and Royce lobbied Al to have Scott work for Al versus letting Al bring in his own salesperson. Al had the final decision.

They could have chosen to leave Scott in the circuit-sales MFS organization as the City Director of Houston. He had the largest financial incentive to have Houston succeed versus all of MFS since his stock ownership was in MFS of Houston, not MFS Corporate. All this came up in the interview with Al Fenn. The VP of Sales job under Al, Scott was convinced, would yield terrific results for every city including Houston, ultimately benefiting all the stock, his included. He was right.

Security and the Internet vs. Circuits

 The first half of 1991 was spent developing and writing the Houston Data Services business plan, which defined *"Transparent LAN Services"* and described the concept of a closed user group over shared FDDI using encapsulation.[102]

Today these same issues occur with respect to the use of the Internet. You can build a private network on circuits. A virtual private network can be built on a statistically multiplexed backbone such as frame relay or ATM, but using a shared IP network called the Internet requires a different form of security.

Using the Internet in a secure way requires putting firewalls at every site and even these provide only limited protection. The level of security necessary may require every desktop and server encrypt all files and messages. That was impossible in the early 1990s.

---

[102] http://tinyurl.com/RoutingPaths-BusPlan

Multiplexed circuits are implicitly secure because a fixed time slot is dedicated, not shared. There is no co-mingling of the data at all. Meanwhile packetized traffic on a shared medium using a closed user group is considered insecure if the shared medium is LAN in nature rather than circuits. The data is commingled, thus potentially visible. The use of closed user groups in Frame Relay or ATM using the concept of Permanent Virtual Circuits (PVCs) is perceived to be secure in the same way that multiplexed circuits are, and considered more secure than encapsulation over a shared medium. However, the data still flows unencrypted on a shared media, thus is less secure, a distinction often lost.

The Internet is totally open but it can be made more secure with firewalls and encryption. Adding encapsulation, or tunneling over IP, using an encrypted tunneling protocol such as *Internet Protocol Security* (IPsec)[103] with additional application layer encryption using *Transport Layer Security* (TLS)[104] is considered the secure way of using IP over the Internet. Modern, powerful encryption capabilities provide encrypted tunnels which can be quite secure within a shared media, but such capabilities did not exist in the 1990s. RFC4301[105] defining modern IPsec was published in December 2005, though earlier versions existed.

The term *Metropolitan Multimegabit Data Service* (MMDS) was chosen to describe the MFS transparent LAN service. Tim Devine coined the term as a joke to play off the RBOC jaw-breaker, *Switched Multi-megabit Data Services* (SMDS).[106] SMDS was the RBOC approach to networking in a metropolitan area at speeds higher than T1s. Everyone chuckled at the name. MFS was offering a service that worked for high-speed LAN interconnection but did not require the customer to connect with a new protocol. Meanwhile, SMDS defined a new protocol that would have to be used by the company connecting to the RBOC network.

This drove the RBOCs nuts because their approach required millions per city to turn up the first customer while our approach was success based so the company did not spend money until they received an order. That is why Royce was willing to gamble on setting up the Board meeting. The concept was one of using technology that was considered customer premises equipment in MFS points of presence. The equipment was only installed upon receiving an order. This allowed the company to offer the service in the market without buying equipment before receiving orders.

Michael O'Dell was at Bell Labs (now Nokia Bell Labs, owned by Nokia) at the time. Stan Hanks told him about the LAN service. He stated that the thing that scared the RBOCs the most was that an end user could understand how to use the MFS service in less than 30 seconds. Yet it would take hours for an end user to understand Frame Relay or SMDS. The RBOC penchant for jaw-breaking nomenclature was often grist for humor.

---

[103] https://en.wikipedia.org/wiki/IPsec

[104] https://en.wikipedia.org/wiki/Transport_Layer_Security

[105] https://tools.ietf.org/html/rfc4301

[106] https://en.wikipedia.org/wiki/Switched_Multi-megabit_Data_Service

Mike later became the CTO of UUNET and in 2002 became a Venture Partner[107] at *New Enterprise Associates* (NEA).[108] NEA was also an early investor in UUNET.

MFS discovered there was a great deal of interest in high-speed connectivity in NY. Scott and his team started spending the latter part of 1991 in New York and DC. Consequently, they did not spend time in Houston, and sales languished.

MFS also sold a few orders for high-speed connectivity around the country before MFS Datanet was launched. The latter half of 1991 was spent helping the MFS sales force. The team created opportunities that led to orders. Scott worked on a Metropolitan Area Ethernet (which eventually became MAE-East) with Rick Adams at UUNET, Suranet, and PSI. They developed solutions for CNBC and wrote the specification for the NASA FDDI/Channel Extension deal in DC. These turned into real orders in 1992.

Operationally, MFS had several people who made a difference early on. Dan Strange out of Chicago was a great engineer and hard worker. He spent a tremendous amount of time understanding the details of engineering the services and evaluating the vendors. Dan figured out how they could exceed the 100km distance limit of FDDI using fewer nodes. Still, they could not exceed the total Token Rotation Time in the specification.

No one had ever attempted to exceed the 100km specification of FDDI. MFS had an 80-mile ring that, when put in full-wrap, would extend almost 3 times the distance allowed by the specification. The optical budget was not a problem; Scott used his fiber optics systems design expertise to convince people not to worry about that. However, the Total Token Rotation time issue implied that the speed of light must increase to accomplish the feat of going farther than 100km (approximately 60 miles).

As a joke, Scott claimed MFS engineers could increase the speed of light by changing the gravitational constant of the universe, but could not get FCC and OSHA approvals. (There's a core of truth behind the joke. In glass fibers, the speed of light is almost a full one-third slower than the speed of light in a vacuum. To extend the FDDI Rings beyond the 100 KM limit it is necessary is to speed up light in the fiber.) In 2013 Researchers at the University of Southampton in England produced optical fibers that can transfer data at 99.7% of the speed of light in a vacuum. If an FDDI Ring were built using this fiber, that 60-mile limit would be significantly raised. However, FDDI was rendered obsolete in 1995 when Fast Ethernet was introduced. It is intriguing the way technical jests often melt away under the relentless pace of advancing technology.

John Calhoun and Roland Freund helped install and test numerous vendors' products. During the evaluation phase of this project, we were looking for a workable vendor solution. Their combined background of knowing the transmission equipment and learning the LAN side of the equation was very useful.

---

[107] http://www.nea.com/team/mike-odell
[108] https://en.wikipedia.org/wiki/New_Enterprise_Associates

One day in the first quarter of 1991 MFS sought to demonstrate video over Ethernet for the *Houston Chronicle*. They needed to confirm that the FiberMux Magnums worked over a DS3, as all previous installations had used dark fiber. One morning Roland and John unboxed some Magnum units provided by Mike Schmitt, the local distributor of FiberMux, and installed them on a DS3. Everything worked as expected, and they easily made Ethernet operational over the DS3. The video-over-Ethernet startup company out of Dallas came to Houston and their installation on the system demonstrated their technology impressively.[109]

Thus in early 1991, MFS demonstrated compressed TCP/IP video over Ethernet, up and running in a metropolitan area. MFS deployed the first commercial version of IP working between buildings over a fiber optic transparent LAN service from a network provided *Competitive Access Provider* (CAP).

Scott took the ease of installation, and the comments of Roland and John, back to MFS at Oakbrook. They were part of the ammunition needed to prove that the technology worked and that, with training, MFS regular technicians could install it. Roland and John could demonstrate that this was practical. This overcame a major issue to the operations people in Oakbrook.

A corporate issue preventing the rollout of Data services was overcome by the successful efforts of two local operations people. When Scott presented to the Board in April he noted that an operational Ethernet link on the DS3 network had been installed in one morning by normal technicians. John and Roland were in fact exceptional people; Scott's point was that they were not highly-trained data technicians.

The concept of selling a national data service offering that included the backbone and local loops required a major shift in the overall scope of the MFS business plan, and implied commitment to compete for some services with its biggest customers, the *Interexchange Carriers* (IXCs). The decision was monumental, and was opposed by numerous people within MFS who did not believe in the future of data over voice. No one then could recognize the impending death of the traditional switched voice hierarchy and the upheaval the next few years would bring.

Jim and Royce had to address this opposition at MFS. Royce came to Houston once Al Fenn was on board and thanked Scott "for dragging us kicking and screaming into the data networking business." The willingness of Royce and Jim Crowe to change their minds underscored the depth of Al Fenn's leadership abilities. He understood creating and selling data services in ways traditional telecom people did not.

Al Fenn and his team wrote a business plan for MFS Datanet with Scott's valuable input. The first sentence in the business plan states that Datanet was going to "build a business... unlike any... telecommunications business." Again, it was Fenn who convinced the Board to invest in MFS Datanet, a radical new business at that time. In mid-1992 MFS Datanet launched with a core team of people who followed Al Fenn from BT Tymnet.

---

[109] MFS FiberMux Press Conference 1991 - https://youtu.be/ull-rM1TXxs

The people who Al pulled together in San Jose to create the new business were exceptional, and it gives me no end of personal pride and satisfaction to have been among them. The core group was multi-talented and experienced. They developed the key capabilities needed to deploy the national backbone. For instance, Bill Euske and his group were critical. They provided MFS Datanet with the ability to create and deploy services in a new and unique way. Bill's open-minded approach to using technology and his experience made him unique in the industry.

The team Bill Euske pulled together (consisting of myself, BJ Chang, Dan Lasater, Ron Whitlock and Steve Feldman) is the best group of Data Engineering people who had ever existed in one carrier, and a team of people I am proud to have known and worked with. They knew how to utilize traditional telecommunications platforms and data products to create new services in the industry.

Ken Holcomb is versatile and his strengths and abilities were essential in the early stages of development of the company. His "utility player" capabilities allowed him to deploy the national backbone while installing the existing services. Some of his key people were Tommy Waldrop, and John Calhoun out of Houston where the data services started. Brian Roberts who was out of D.C. came on the scene later and bailed us out of several problems. Rodney Elder was instrumental in getting orders and helping get them up and running.

A National ATM Backbone

Ken Holcomb was responsible for getting the initial internal national LAN backbone operational for MFS Datanet employees. He was also responsible for building the Network Control Center. We were eating our own cooking. When you went anywhere to an MFS city and were in an "On Net" building you could log on to the servers in San Jose at native LAN speeds and work from that city. That was the single largest Microsoft "Windows for Workgroups" national LAN in the world at that time. It may have been the largest single non-routed IPX and NetBEUI bridged LAN backbone running multiple protocols ever. Windows for Workgroups 3.1[110] was an "add on" package to Windows 3.1 released in October 1992. It let a PC function as a peer-to-peer networking device so you could share files between PCs directly from their C drive without going to a server. It also allowed you to drop files onto a server or pick them up from a server directly if the server was shared. It used the Microsoft non-routable protocol[111] *NetBIOS Extended User Interface* (NetBEUI)[112] while the Enterprise ran on Novell IPX protocol for client server applications including Microsoft Mail.

This was a big deal, and important to the sales force for big files like Freelance Presentations (a Lotus product), which they were using in lieu of PowerPoint. Once Ken got the backbone functioning, he took over operations for getting orders installed. Overseeing the internal LAN and National Ethernet LAN backbone with the applications running over it was a big job but MFS itself was using the national backbone product it was selling. First-hand experience informed every demo showing customers what was possible with the native Ethernet backbone coast to coast.

Tommy Waldrop John Calhoun and Rodney Elder were instrumental in teaching San Jose (and all operations people) how to install the services using the FiberMux gear. Brian Roberts came on board in mid-1992 as the operations manager in Washington D.C. He learned under fire how to install MFS Datanet Services on top of the existing fiber infrastructure of MFS. Their work in 1992 was critical to MFS to delivery of any products or services in the early stages of the business, and cemented credibility with the big customers recruited as early adopters.

SunExpert Magazine (later *Server/Workstation Expert Magazine*, now no longer published) ran an article[113] written by Stan Hanks on MFS Datanet in October 1992. He described how MFS was changing the face of the industry by offering new services different from any other players, and a radical shift from traditional phone company services. He notes

It is interesting how Stan Hanks explains the other options, such as point-to-point circuits, Frame Relay or SMDS, which sounded good in theory but required a huge capital outlay by the RBOCs. Bringing the first customer online was extremely expensive in most cases. RBOCs were slow and reluctant to invest.

---

[110] https://en.wikipedia.org/wiki/Windows_3.1x - Windows_for_Workgroups

[111] http://www.pcmag.com/encyclopedia/term/48055/non-routable-protocol

[112] https://en.wikipedia.org/wiki/NetBIOS

[113] http://tinyurl.com/SunExpert-Bottleneck

MFS Datanet also offered Channel Extension services (34 Mbps), a proprietary IBM host mainframe service that formerly had only been available via dedicated copper cables. Bill and Jay at Centel who noted MFS could offer this service using the Network Systems fiber optic version of Channel Extension.

It is also interesting that Stan mentions Atlanta was just being turned up in late 1992. One of the key people who became part of the Atlanta sales organization was Larissa Herda,[114] who later went to Time-Warner.[115] Scott spent some time in Atlanta with her and others teaching them about the standard MFS circuit products, how they were provisioned and how LAN Data services worked on top of this infrastructure. He showed the MFS sales people how the circuit based services worked in conjunction with the MFS Datanet services and since he had been the City Director for Houston and an early employee of MFS it was useful for him to explain how these services worked together.

X.25 or Layer 2?

During the early planning phases of MFS Datanet in 1992 Scott discussed with Al Fenn, Ken Holcomb, Bill Euske and Jay Jonakait the business focus of high-speed LAN services as he had envisioned, or whether Datanet should go after the largest installed base of computers, which was IBM and use X.25 as the network? Scott recalls that with the history of Tymnet-style networking, X.25 seemed a more natural choice at the time. Scott spent time educating Al and his team about the way LANs worked in a campus environment and how we wanted to make them work across town or across the USA just like they worked in the riser of a building.

Creating a Layer 2-bridged LAN environment that was "protocol agnostic" was a radical idea at the time. Had we chosen a single protocol, no doubt we would have chosen X.25, which in hindsight would have been a monumental mistake.

With all due respect to Scott, I think the focus on X.25 was overblown. Yes, it was considered, but TCP/IP was coming on strong and Internet-style networking was rapidly gaining ground. I vividly recall Al Fenn remarking that the Ethernet LAN connector was the "RS-232 port of the 1990s" when discussing how we would interface to the customer. Scott says some called it crazy that the protocol-agnostic environment won out. In hindsight, modern designers would insist it should have been routed TCP/IP, but that was a different world. By staying protocol agnostic, we could carry TCP/IP when it became important. I do not recall more than passing consideration given to building an X.25-centric network, but the battle between Layer 2, protocol-agnostic architecture and a routed Layer 3 model was a different debate and one that continued for a long time.

---

[114] Larissa Herda became the President and CEO of Time-Warner Telecom June 19, 1998. She grew it into a huge company selling dedicated circuits, switched voice services, data services including Ethernet services, and of course Internet services. They were acquired November 1, 2014 by Level 3 Communications for $5.7 billion. Larissa is extremely savvy in her ability to evolve the services offered by Time Warner Telecom. This is especially true considering the huge changes that have happened in the Telecom/Internet Services industry since the early 1990s and the high tech/Internet bubble stock crash that occurred in April 2000.

[115] https://en.wikipedia.org/wiki/TW_Telecom

Scott explained what he learned from the focus groups: that we wanted to pass all the protocols through our connections. This included the non-routable protocols. He spent a lot of time explaining the things learned from the Houston Focus Group sessions.

Scott says that eventually after heavy debate and discussion, Al and his team agreed we would be leaders in developing the next type of solution versus just another X.25 data network. I was not privy to Al's early thinking, but at the time I joined in mid-1992, X.25 was not on the table in any form, not even as a side-dish.

Sell, sell and re-sell your ideas.

Frame Relay was a relatively high-speed new technology that went from 56 kbps up to about 4 Mbps. A newly-installed base of data services called "Frame Relay" was a statistically multiplexed data service offering that Sprint and Wiltel were pushing in 1991-92. Scott was opposed to building just another data service that required cards on client routers to convert the LAN protocol to the carrier protocol, partly because it had to be engineered to set up "permanent virtual circuits" or PVCs that direct traffic to the location the customer was interested in reaching. It did not operate at native LAN speeds, and other carriers were oversubscribing their backbones, risking congestion by subscribing customers. Scott felt strongly this was an inferior service, difficult for the customer to engineer a network around even if not oversubscribed. The engineering decision had been made by Sprint and WilTel when they offered it to the market.

Scott was talking to Enterprise customers who had been in the Houston focus group as well as all other MFS cities while keeping in touch with startup private IP backbones like UUNET. Guy Almes of the original focus group had moved from Rice University Sesquinet (a Houston-based Internet provider, acquired in Dec 1997 by Verio Inc./On Ramp) to a startup formed in 1990 by NSF Partners Merit, IBM, and MCI called *Advanced Network and Services* (ANS).[116] At the time, ANS was attempting to sell connections to their backbone to research-oriented Enterprise customers who would meet the acceptable use policy requirements of the NSF. ANS was hoping to be chosen "Commercial Internet backbone" at some point in the future and many people thought thcy would be because of their pedigree. Had that happened, the Internet of today may well have been stunted, or even stillborn, controlled by a single monopoly-like entity with little competition.

---

[116] https://en.wikipedia.org/wiki/Advanced_Network_and_Services

Scott was also collaborating with Rick Adams who founded and controlled UUNET, soliciting the perspectives of key players in the industry, input from Enterprise customers and new private IP backbone companies. He was hearing that the Enterprise wanted "closed user group" private multiprotocol LAN backbones more than they needed a technology like Frame Relay. They preferred the service to be more deterministic and not oversubscribed to keep their services from being impacted by others on the same system. Plus, they wanted direct control of any oversubscription themselves. Services sold this way cost more money and reduced margins, and would make things more difficult for ANS since none of the other players were offering services this way. But Scott was encouraged to avoid the Frame Relay quagmire, difficult for us since none of the other players were offering services this way then.

Scott believed in getting "early adopter customers" on the line that would commit to certain types of services once the national network was built. To him, this was a very important part of making sure the right things were being done in the eyes of the customers. Scott was singularly focused on delivering the types of services that customers had told him they wanted, not things like Frame Relay and X.25 the industry, or "experts" said were needed.

In the summer of 1992, Scott was focused on talking to customers while Al and his team in San Jose were writing the business plan in the summer of 1992. Al wanted Scott of help finish the plan instead. Al's request was reasonable, but Scott chafed under it. At the time, he believed he'd be reselling the Transparent LAN Services concept versus the X.25 Data services as a backbone.

The process produced heated discussions with Al and Ken about the best way to set up and provision the business so customers would get what they wanted. Al, Ken, and Bill were focused on making the product repeatable, scalable, and manageable. They were also focused on what had to happen to make sure the network stayed up and was reliable.

Scott knew reliability was essential, mindful of the old MFS Telecom days where the service had to stay up or the customers would not buy the service. He was reminded of the CECO compressor days where loss of a compressor would cause an entire gas pipeline to go down, and money was lost every minute the pipeline was down. Data networks pumped information, and this information was money as surely as gas or oil. Downtime was expensive and would not be tolerated by customers. They agreed whole-heartedly that network architecture and hardware chosen had to deal with failure of parts.

Tradeoffs had to be made. Scott convinced them that native LAN speeds of 10 Mbps for Ethernet and 16 Mbps for Token Ring were essential to the customers who wanted LAN connectivity. Other customers wanted native LAN interfaces but who could not afford the cost, so there was a need for fractional speed LAN connectivity. These discussions influenced the technology decisions, creating a point of contention between Stan Hanks and Al Fenn and his team. Al decided to pick a new technology called *Asynchronous Transfer Mode* or ATM.

Stan did not agree. He wanted to stay with bridged technologies that used TCP/IP networking at that time. Once ATM was chosen, he stopped being a consultant; MFS Datanet moved forward without his help after that.

Al Fenn's choice of ATM engendered a risk based on the need to put multiple customers' traffic on one backbone like a DS3 and later OC3 while keeping customers' traffic separate and secure. MFS needed to create a "closed user group of LAN interfaces" for just one customer, while keeping it separate from another customer's traffic. ATM was the only viable way of accomplishing this in mid-1992. It also allowed MFS to scale up its backbone, go with more diverse routes, go to more cities and eventually have FDDI interfaces as needed. Finally, ATM technology had to keep operations secure, robust and highly reliable for customer throughput regardless of the customer protocols.

Scott trusted Stan Hanks' opinions by then since he had worked with him for over 6 months. He was torn, as he also knew that the CEO/President of the new company MFS Datanet had to make his own decisions and live with them. This was a true leadership decision.

Once Al Fenn and his team decided to go with ATM, Scott got behind it 100 percent. Putting routers on DS3s without creating too much latency and delay was not possible. Routers then were relatively slow, and each router-hop added significant delay compared to an ATM switch. A coast-to-coast backbone would have at least three routers plus the "speed of light" delay. The company's ability to disagree; then reach, understand and appreciate a final decision was something Scott knew didn't happen every day.

Explanations shifted from the "IP tunneling closed user group" construct for the Metro FDDI services to showing that Datanet was deploying the world's first commercial ATM solution. The TCP/IP community from the early ISPs did not like ATM, which required breaking IP packets into ATM cells, transporting them and reassembling them on the other end. This was inefficient in their view and counter to the purpose of using TCP/IP in any transmission service. We explained how much flexibility it gave us as a service provider.

Waiting on delivery of the first ATM switches was a challenge at the end of 1992. Early adopters like UUNET and private Enterprise customers were critical to getting the business up and running. Some of the customers who wanted Metro Ethernet or Token Ring just used the FiberMux TDM solution of Ethernet TDM over DS3s as a stopgap while we waited on the ATM vendors. Thanks to TDM, there was no contention between channels, and it was easy to understand and install

Initially there were no plans for a dedicated data services sales staff. MFS Datanet would use the MFS telecom sales force. Working with marketing, we had to create training materials and presentations for training the sales force to sell to customers and create MFS Datanet brochures.

The cover of the first Datanet brochure[117] came from discussions with Head of Marketing Bob Barbour, and Stan Hanks as a consultant for MFS Datanet, in Scott's living room in Houston. They decided to establish a metropolitan and national concept in one picture. MFS was still very much a "metropolitan-only" company so the brochure had to illustrate the national concept as well.

---

[117] http://tinyurl.com/RoutingPaths-Connectivity

The HLI white brochure was developed as a handout and was used as the boilerplate of many MFS Datanet proposals. It describes the services as they were originally envisioned. Its graphics illustrate how the fiber runs under the streets and connects buildings within a city and then between cities.

An April 1992 internal memo discussed the fact MFS must be willing to use any technology, referencing a future where we would use ATM instead of the TCP/IP encapsulation scheme. We had designed an IP encapsulation approach that created closed user groups over a shared transport like FDDI. Clearly, we had not chosen ATM officially in April of 1992. The memo defined how to create a market and drive a new paradigm in the market. These concepts were intended to influence the company and keep MFS from being perceived as "just a phone company."

 The June memo[118] from Scott to the City Directors and their managers clearly laid out sales strategy: telecom salespeople were needed to sell these services. Scott had convinced Al of the importance of technical sales consultants to help sell these data services to the customers. He came up with the name Network Communications Consultants because that was going to be their role in the organization. We called them NCC's or Network Communications Consultants. The actual name was initially suggested to Scott by Ben Gerenstein while he was in the process of being interviewed for the new position. Scott liked it and used it.

 The role of the *Network Communications Consultant* (NCC) was defined in connection with the role of the telecom sales representative. A memo documents how we made a serious attempt to indoctrinate the telecom sales force with the correct ideas and attitudes behind Transparent LAN Services. Scott spent a considerable amount of time defining the training concepts, creating the slides,[119] and conducting classes in all the major cities. Meanwhile, he also made joint sales calls with the sales representatives, training them by working closely with them in person on real sales opportunities.

All this required weekly trips from Houston to all the MFS cities: New York, Boston, Baltimore, Philadelphia, DC, Atlanta, Chicago, Minneapolis, Dallas, Los Angeles, San Francisco and San Jose. He enjoyed the work and was the only person inside MFS routinely involved in both sales and training. Coupled with his early-industry background, his experiences bestowed perspective shared by few others in MFS.

 One of the early adopters approached in early 1992 was Ken Starkey, the head of all the private networks at Bear Stearns. Scott had been explaining the idea for Ethernet or Token Ring Transparent LAN Services[120] to which Stan Hanks added the IP encapsulation over FDDI service concepts. Ken liked this approach but preferred the TDM FiberMux version for his own metro Ethernet links. His input underscored that customers felt, correctly or not, that ATM and TDM were more secure.

---

[118] http://tinyurl.com/RoutingPaths-June92
[119] http://tinyurl.com/RoutingPaths-Customers
[120] https://youtu.be/QPeqcksKG6A

Ken had told Scott about MPR Teltech Ltd., an unknown company among vendors. Ken said MPR had an ATM switch that could adapt Ethernet to ATM on a DS3. This was great news since Al and his team wanted to go that way and the industry was becoming very interested in the potential of using ATM for high-speed data networking over fiber optic circuits. Scott, Dana Crowne, Phil Hamlin, Bill Euske, Ken Holcomb and I all flew to British Columbia and looked at the equipment that MPR had working.

Bill Euske and Ken Holcomb liked the equipment well enough to recommend it to Al Fenn. We had to deploy technology for the national backbone very early in the history of ATM. Newbridge, which made TDM T1 multiplexers acquired an exclusive license to MPR's ATM technology in 1992, just as we were selecting them. This meant deliveries on the products were slowed. The decision to use ATM and MPR Teltech Ltd was made sometime in mid-1992. Newbridge went on to buy MPR's entire ATM business unit, and then France's Alcatel bought Newbridge in February 2000 for $7.1 billion, citing ATM technology as the primary target. MFS Datanet also used GDC Apex ATM switches in the second generation of services.

Using the TDM solution as a stopgap solution was dicey when talking to customers about delivery of national services. We were selling while still choosing vendors and then booking orders while deploying technology that was being used for the first time. The risk was not something AT&T, MCI, or Sprint—and certainly not the RBOCs — would have taken. We told the early adopter customers the truth about this. They respected that and what we were doing, and worked with us as we sorted things out on the fly. Early adopters want to know their risks and have input, which is part of why they become early adopters. They are part of the process.

We also had to figure out how to turn up services and manage them using this new technology. No one had ever used ATM as a service offering, nor was anyone also trying to use Ethernet adaptation to ATM as the definition of a metro and long-haul service. This was unique and much more difficult than deploying native ATM hand-offs to the customers.

Al Fenn made the decision to use ATM knowing there were tremendous risks. He felt that the low latency (1 millisecond or less delay through the switch while routers were well over 100 milliseconds) and scalability (plug lots of circuit cards like DS3 and later OC3 or lots of Ethernet ports on the switch) of the technology was worth the gamble.

The MFS ATM brochure explains the benefits of ATM for data applications as we saw them at that time. Al could see the inherent benefits to selecting ATM as the backbone technology. He gambled in a big way to get it deployed, which delayed offering national high-speed LAN connectivity, even though we had some parts of the country operational before the end of the year — another of the many geographically limiting issues sales had to deal with.

Since these services differed from what was available at the time, we had to create slides to illustrate what customers would be buying. Connecting to a public network using the physical interface of the LAN was a concept new to the entire industry. Illustrating an Ethernet, Token Ring or FDDI connectors inside a building, layered on top of the MFS telecom services, came from discussions with Bob Barbour in San Jose.

Bob created that slide using Lotus software called Freelance. He scanned some old MFS literature in for the buildings and fiber in the streets; then created a drawing that has been used to illustrate the transparent LAN concept very clearly to customers ever since. Offering a service that was not a circuit essentially changed the industry.

For the first time a "phone company" would meet the customer with the physical interface and protocol that the customer wanted. Our model didn't force the user to meet the public network with a special protocol. We didn't assume the "long haul" network was the center of the universe. This was Scott's original idea, endorsed in focus group meetings and now confirmed by real-world customers in the marketplace.

The edge of the carrier network where the customer plugged in was all that was important to the end user Enterprise company. The middle became transparent to the customers. That was the whole point: the customer did not have to worry about what the public service provider used as a technology.

 This is completely different from the concept of managed services, where a customer picks vendors and technologies and takes on the burden of vendor selection for everything. MFS eliminated the need to engineer those issues. We developed another simpler representation[121] of this as well.

Local metro high-speed data services were generating increasing interest across the country. We had our first installation in Chicago for the Natural History Museum in 1991. Rodney Elder was instrumental in making that happen.

The Museum needed to interconnect two locations with Ethernet that were across the street from each other and several thousand feet apart. This required right of way access. We ran the fibers and installed the FiberMux TDM boxes, creating 2 Ethernet channels between the buildings, which allowed two computer networks in two buildings thousands of feet apart to work as though they were in one building. The home of MFS Corporate had its first operating customer.

Financial institutions in Chicago, New York and San Francisco were early adopters of the transparent LAN technology. High-speed, low latency access between computers was important to making fast decisions about stock trades and executing on them faster than the competition. The analytic engines in a central location, with high-speed Ethernet feeds coming from the computers that executed the trades for the exchanges in other buildings across town. Mission-critical to these companies' abilities to make money were fast connections with "low latency or low delay."

When Royce Holland and Jim Crowe decided to kick it up a notch and to start a new business that would only focus on High-Speed Data Networking, it was not a service that was competitive with long-distance carriers and was not a big part of the industry at that time.

---

[121] http://tinyurl.com/RoutingPaths-Transparent

Long-distance carriers were MFS Telecom's biggest customers at the time and we did not wish to anger them by creating the appearance of competing for the same customers. None of the carriers like MCI, AT&T, Sprint, WCOM or others had a dedicated private data backbone set up to only sell high-speed data services. We thought we could do this and not compete with, or even be noticed by the long-distance companies.

Something Called The Internet

Scott first became aware of the Internet when he met with Stan Barber, of the *Baylor College of Medicine* to explore bringing fiber to the *Houston Medical Center*. Barber was knowledgeable and eager to collaborate, and he connected Scott with Guy Almes at Rice University. Almes was interested in high-speed, high-bandwidth metropolitan-wide service for the University Internet connectivity. This was Scott's introduction to the NSFNet academic research network. Stan Barber and Guy Almes ran Sesquinet, the Texas Regional Information Service Provider.

The NSFNET had an *Acceptable Uses Policy* (AUP) requirement. Only research institutions; universities and companies with a research department could get connections and conduct research on the Internet. No commercial uses were permitted. Anyone not connected to a research institution could not buy an Internet connection. Few people outside of that research and government community knew about the Internet, understood it or cared about it.

Almes and Barber introduced Scott to ex-Rice University Ph.D. student Stan Hanks, who was also a consultant with *Technology Transfer Associates* (TTA). Scott understood the Metro Fiber business and Layer 2 LAN connectivity via fiber, but little about TCP/IP. Hanks lacked Scott's metropolitan connectivity expertise but was well-versed in TCP/IP and the application layers.

Selling hardware and equipment as a systems integrator was very different than developing a business offering a service for a monthly fee. Scott had developed TDM-based services over fiber as NCI and now Metropolitan Fiber Systems local loop services. He had envisioned such Metro LAN connectivity over Fiber for years.

Many types of expertise and skills were needed to explore the details and to identify all related issues and pitfalls: a sophisticated gang of collaborators, experts who lived and died to keep their networks up for employees and customers. The people Scott began gathering understood the proposition was a good idea, and began collectively to identify precisely what "it" was.

In defining the performance and interconnection requirements from various perspectives, they had to be particularly clear about the costs and revenues. No matter how good the idea seemed, unless it was profitable, it would never happen.

Scott could draw people from companies like "Big Five" accounting firms such as the storied Arthur Andersen[122] to determine what customers wanted and to propose solutions as services. Scott was excited, and humbled by the willingness of these professionals to participate. They knew real-world networking issues. And here they were, eager to be honest, objective and helpful in discussions with MFS.

 Scott dragooned one of the sales reps into videotaping a focus group in action using his personal VCR. The quality is poor; the shots are often not focused on the speaker. The whiteboard is almost impossible to read. But the discussions are audible and interesting. Customer driven[123] services at MFS were developed using real customer feedback.

---

[122] https://en.wikipedia.org/wiki/Arthur_Andersen
[123] https://youtu.be/C0D0Tl642rM

Bill McCuistion was an entrepreneurial-minded expert for Arthur Andersen, based in Houston. He dealt with the key network technology people in their Chicago headquarters, so his input represented views of key network people in a sophisticated large Enterprise. His insight was distinct from others and raised different issues than other companies. His input was critical.

Groups discussed the issues[124] associated with interconnecting offices within one Enterprise and how this can be different from an Enterprise interconnecting to its clients in the manner Arthur Andersen was seeking. Then Guy Almes and others discuss how the Internet is "any to any" and that is yet again another type of service and has its own set of issues.

Scott's original focus was on how best to deploy Private Network services to interconnect multiple locations within one Enterprise. The video shows how these conversations with our customers opened our eyes to additional services we might create. Other interconnection needs, including groups of customers interconnecting privately were explored, raising the idea of multiple companies interconnecting via shared network resources similar to what the government-owned Internet offered.

The Internet was not commercialized at the time of these videos and we did not discuss or focus on how we could create service offerings to "commercialize" it. The concept of multiple companies interconnecting using routers was discussed but Scott feared MFS was not up to managing routers. We barely had anyone who knew what a Local Area Network like Ethernet was, so adding routing to the equation was too much for MFS to absorb in 1991.

Developing the Internet Plan

During the late-1991 to early-1992 time-frame, Scott was heading up the high-speed data networking sales efforts, pushing to sell to the Enterprise as an "on-net" customer. He was also learning about the new concept called the Internet, which he examined as a new type of customer for MFS local loop circuits and for the MFS Datanet High-Speed LAN services. This was a potential source of good customers, even if the industry was new, uncertain and underfunded.

In 1991 Scott went with Stan Hanks to Interop, the biggest industry trade show for networking equipment. This trade show for all networking devices included Statistical Multiplexers, Frame Relay switches (Layer 2) or TCP/IP Routers which operated at Layer 3 of the OSI[125] model, and Bridges which operated Layer 2 of the 7-layer stack. This trade show focused on demonstrating interoperability of this wide-ranging, often early-stage equipment.

We often speak of the OSI model, and in those days OSI was given a lot of credence. TCP/IP networking today does not concern itself as strongly with strict hierarchical encapsulation and layering. The interested student is invited to further explore the RFC 3496 section "Layering Considered Harmful"[126] for more info. Nonetheless, the OSI model remains generally useful to distinguish these functions.

---

[124] https://youtu.be/eVl2BxEMAjQ

[125] https://en.wikipedia.org/wiki/OSI_model

[126] https://tools.ietf.org/html/rfc3439 - section-3

At Interop, vendors were showing products suitable for provisioning the metro high-speed transparent LAN services, then officially called Multi-Megabit Data Services (MMDS). Here Stan introduced Scott to Rick Adams, founder of a company called UUNET/AlterNet, a provider of "commercial" Internet services. This proved a momentous introduction. Founded in 1987, UUNET/AlterNet was among the earliest ISPs, along with PSINet.[127] MFS worked with and sold to both companies.

Rick announced he was the leader who was going to move the Internet from a government-funded academic resource into a viable commercial business model, and not let the NSF define it and bestow it on a monopolistic-minded carrier like MCI or AT&T. Scott, new to Rick's "Internet" ideas, just listened and learned. He did not immediately decide whether this was going to be true. Adams might succeed or fail, but Scott felt Rick had the potential to be a big customer. He wanted to do anything possible to help him develop his business using MFS services. Scott knew from Stan Hanks that Rick was influential in this new growing group of companies then called private IP providers, and helping him succeed meant MFS Datanet would be developing a unique set of new customers.

The language was evolving. The term *Internet Service Provider* was not yet being used. The ISP acronym was being used in a very different way in late 1991, for content players like Quotron, Reuters or Bloomberg. These *Information Service Providers* (ISP) had huge potential to use this new data service, as well as private IP backbones (later known as ISPs) and Enterprise private networks. Today the acronym ISP only means Internet Service Provider; the others are called Content Providers and those selling news and stock analytics are still called Information Service Providers, but no longer use the acronym ISP.

Scott was very interested in selling Rick Adams LAN derived services between his locations in the D.C. area. Rick had big plans that made him a likely customer for a national Ethernet backbone between the locations he sought to deploy across the US. The concept of a national backbone owned and operated by a private IP network operator independent of the NSF backbone was the beginning of the idea of a "Commercial Internet," routing traffic over a backbone that did not have a restrictive acceptable usage policy and was not funded and controlled by the federal government. Rick introduced that idea to Scott, who instantly recognized the merit. His instinctual assessment was then reinforced by the Focus Group meetings and since MFS was Layer 2 Bridged, and Rick and others were Layer 3 Routed, it fit well within the scope of the planned transparent LAN services Scott was rolling out.

Commercial use of the Internet was a new and radical concept. Rick aimed to be the founder of a commercial Internet backbone. Based on those meetings, Rick Adams was the creative driver who strategized the interconnection of these early Private IP networks.

[127] https://en.wikipedia.org/wiki/PSINet

Dial-up and a Flood of Minutes

Rick Adams is credited with creating the Serial Line Interface Protocol (SLIP), the connection protocol initially used by all dial-up users of the Internet. His 1984 implementation in Berkeley Unix, and for Sun Microsystems workstations, was released freely to the world and became a standard dial-up methodology for more than a decade until replaced by PPP.

SLIP[128] is depicted used in the movie "You've got Mail" but that was an AOL connection, not an Internet connection. SLIP became a de facto standard method to transport TCP/IP over serial lines but was never blessed as an official Internet standard. PPP[129] would replace SLIP as an official Internet standard solution. Rick was already famous in Internet circles, for his work on BSD Unix, SLIP and PPP, and for his work supporting users accessing the NSFNet Internet when he and Scott met.

AOL was not the Internet but a private backbone and content / information service provider, a distinction few understood. MFS sold thousands of lines and local loops to them and understood they were a proprietary service. Later AOL crafted their structure around the Internet, first by providing a gateway function so AOL users could access Usenet. They were an early Content Provider before the Commercial Internet captured the public eye, just like CompuServe and a few others. Widely distributed free floppies and CDs contained the SLIP / PPP software to connect to their servers.

Dial up was an important enabler, allowing companies and individuals to use the Internet cheaply. SLIP (and later PPP) enabled users to connect using POTS telephone lines versus expensive dedicated private lines. At that time, an end user dialed a modem bank, which aggregated many callers into a higher speed trunk to reach the Internet. Only the regional research networks or large corporations could afford dedicated lines. SLIP became a critical enabler, allowing the public to connect cheaply. Few recognize how important Rick Adams is to the commercialization of the Internet.

Scott relates how he first joined the Internet community as a user of UUNET, with a real "Internet Email" address, including a floppy disk with the SLIP software. But there was no local number he could call to get online. He must dial long-distance to Washington, D.C. where UUNET's modems were located.

Someone had to supply the modem banks, the local dial-in numbers and the connecting services related to the "dial-up end" of the connection. This was not a trivial thing. A national presence meant more than simply having a backbone.

Tymnet was long a provider of dial-up services, providing local phone numbers where users of the ubiquitous TI Silent 700 terminals could dial-in and connect to Tymnet's network. It should be noted too, that AOL was a major user of Tymnet, and achieved its national dial-in footprint via Tymnet. In those pre-CLEC days, that meant buying services from the phone company and routing large numbers of POTS lines into a Tymnet facility. MFS would repeat that formula.

---

[128] The Serial Line Internet Protocol (SLIP) is an encapsulation of the Internet Protocol designed for serial ports and modem connections. It is documented in RFC 1055.

[129] The Point-to-Point Protocol (PPP) is a Layer 2 protocol used to establish a direct connection between two nodes. It is documented in RFC 1661.

With the establishment of MFS Datanet, the classic MFS business became MFS Telecom, both under the umbrella of MFS Communications Corporation. In these early days, Al Fenn and MFS Corporate were looking for synergies between MFS Datanet and MFS Telecom. Dial-up services emerged as one way the two companies could work together.

MFS Telecom had staked out territory as a provider of services to business and when businesses began having remote users dial-in to their corporate networks, modem banks, and related services, these became one more service MFS Telecom provided. Initially, it was minor side-business and did not have a national footprint. Often such dial-up involved long-distance and the long-distance revenue was welcome by the long-distance carriers.

MFS Telecom installed and managed the banks of dial-in modems and provided a hand-off. When MFS Datanet began pursuing the Internet business and a national footprint, the immensity of the business opportunity became evident.

There was a bit of arcane magic here that would not be obvious to the casual observer. Those dial-up callers were not MFS dial-tone subscribers. Those home Internet users were customers of the ILEC, the local telephone company. A CLEC like MFS interconnects with the local ILEC via a series of interconnection agreements, legal documents that spell out who pays who and how much. Those agreements are far more complex than the technological interface.

In the traditional CLEC/ILEC environment, CLEC customers are businesses, often with clients residing on the ILEC network. After all, most phones by far exist on the ILEC network. When calls originate from a business and terminate at a number on the ILEC network, the CLEC pays the ILEC for each call-minute at a rate dictated by those agreements, that is, each minute, or fraction of a minute the call lasts.

When a home phone user calls a business, the ILEC counts those minutes against the ones flowing the other way. Usually, they tend to balance, with one or the other coughing up a little money at the end of the month. In the traditional business world, often the CLEC pays the ILEC a small amount of money. Either way, it was not considered serious revenue by either party.

With the advent of the dial-up Internet and with the ISPs buying connection services from MFS Datanet and MFS Telecom supplying the modem banks, suddenly, vast numbers of home ILEC subscribers were dialing MFS Telecom modems, connecting to their ISP and staying online for HOURS. The numbers they were dialing belonged to MFS, not the ILEC, and they were all local calls, not long-distance. Those minutes added up, MFS had a hand in the ILEC's pocket.

That strategy of drawing inbound traffic from dial-up users calling the MFS Datanet ISPs via MFS Telecom dial-in modems generated serious revenue.

I attributed that bit of genius to Jim Crowe with input and guidance from Andy Lipman, the legal and regulatory interconnection attorney for MFS, along with Al Fenn's Tymnet experience and I imagine Royce Holland also was a key strategist as well; yet I simply do not know.

What I do know is, I was privy to a couple of conversations about it sometime prior to the sale of MFS. It was impressed upon me that this flow of minutes from the ILEC to MFS Telecom was a secret, extremely sensitive information. The insular and bureaucratic ILECs were relatively unaware, and we wanted to keep it that way.

Napkin Engineering

Stan set up dinner with Rick Adams in 1991 at Interop in San Francisco. Stan explained how Scott had launched his own fiber network and obtained a city-wide franchise in 1987, which let to NCI's acquisition by MFS. Scott thinks Rick was impressed as he considered MFS a big company because they had fiber optic networks installed in eight major cities then. Stan also explained Scott's fiber optic background and how he wanted to create LAN connections over fiber. This led to the focus groups and led MFS to offer native LAN connections within the metro area. They explained to Rick how he might develop a metro Ethernet service to interconnect with his customers or other private IP backbones. Restaurant napkins became critical visual aids as they explored the possibilities.

Scott explained to Rick that MFS was a provider of Layer 2 LAN interconnect services and he had developed these services for any company to use to interconnect locations in a closed user group manner. Rick explained the Internet network worked at Layer 3 and was a routed infrastructure. He liked this distinction because it put him in control and he understood Layer 3 whereas MFS did not.

Scott was not prepared to push a routed service based on the discussions with the focus groups in Houston. It looked like it could be a good partnership. The issues included too much latency delay created by the routers and the need to carry multiple customers traffic over one backbone securely. Routers were for one Enterprise with no separation of traffic. There was also a need to carry non-routable protocols like NetBEUI and DECnet as the Internet only supported TCP/IP.

Rick seemed interested in the concept, pleased that MFS Datanet was not a "phone company." When Scott drew an Ethernet backbone on a napkin and showed how Rick could use router Ethernet ports to interconnect his partners and customers versus using T1s was a breakthrough concept.

Rick pointed out that one T1 connected to each private IP backbone then necessarily had to have all private IP backbones connect to each other one T1 port at a time, multiplying the number of connections required. There would be a lot of T1 connections, more than could be easily supported. With 4 T1s per card and only 4 to 6 cards per router box there was limited real estate available per router.

T1 ports were expensive and low-speed relative to Ethernet. He could only allow these new commercial private IP backbones to interconnect at T1 speeds, T3 being even more outrageously expensive, both for the physical router ports and the cost of the circuit. Rick explained that the formula determining how many T1 ports were needed depended on the number of companies that wanted to interconnect. The number of connections could quickly become immense and unmanageable.

Rick explained the concept of a mesh-connected network where every location connected to every other location. This was the most reliable way to interconnect networks: if one site or node went down then all other locations could still talk to each other, eliminating any single point of failure. This network was expensive to operate, both in cost of circuits and real estate (number of Interface Cards) in the routers. It was a barrier to overcome before the Internet could go commercial.

Rick had been stymied by this math problem because it became too costly and unmanageable to use the mesh network approach to interconnecting the backbones. The more backbones there were the more interconnections. It grows at a function of (n x (n-1) /2) where n = the number of nodes/Interconnected Networks. This formula to calculate the number of circuits or connections shows how unmanageable it becomes and how fast it grows.

Rick explained this by drawing it on a napkin. He illustrated that if you had 3 companies then n = 3, so (3 x 2) = (6 / 2) = 3 connections. If you wanted to make 5 companies or locations connected to each other in a full mesh it was (5 x 4) = (20 / 2) = 10 circuits. So, to go from 3 to 5 players you went from 3 to 10 individual T1. If you went to 15 Private IP backbone companies connected it would be (15 x 14) = 210 divided by 2 = 105 T1 connections, impossible as no router held 105 ports. It also was unmanageable and cost prohibitive. There could be 20, 50 or who knows how many companies wanting to interconnect someday.

This made it financially and physically impractical for all the players to interconnect and end run the NSF backbone. It doesn't scale well and becomes impossible to manage. MFS offered a solution, one that could scale, one that worked, and Ethernet connections are very cheap compared to T1 or T3 router ports. Ethernet was technology companies already used inside of their buildings.

Rick said this issue had been bothering him and other early private IP backbone players for a while. Also, T1 was too slow and too expensive in those quantities. The problem was much worse with DS3 circuits. These circuits used up a lot of resources within the router; the cards were about $30,000 each and the DS3 circuits cost over $3,000 per month per circuit. Rick was right that to try to scale that up is not feasible. End running the NSF backbone and creating a new Commercial Internet was going to be difficult to impossible with this problem looming.

Once Rick understood he could buy an Ethernet connection, a shared Layer 2 medium in which all nodes automatically talk to all other nodes and use only a single Ethernet port on the router, it was like a lightbulb clicked on in his head. Now the small providers could each buy one Ethernet connection and only use half of one card in one router, and all the private IP backbones could be mesh-interconnected easily and cheaply. This was a breakthrough that Rick recognized as such immediately. He appeared excited, shocking to those who knew him.

That launched the discussions about using the MFS Datanet Native LAN Ethernet Services in the Metro Area and we went from there.

The Ethernet connection between all the players solved many problems. This saved expensive real-estate in his router and gave him more ports to sell to paying customers versus connecting to other networks, which was not income-generating.

The Ethernet notion allowed one player to talk to any other player plugged into that shared Ethernet segment without adding physical ports. The idea of a Metro Ethernet made the interconnection of these players physically and economically feasible in a way T1s couldn't match. The value of any-to-any connectivity via a single Ethernet port escalated as more players were added.

MAE IS CONCEIVED

Rick was drawn to the Metropolitan Area Ethernet (MAE) concept. He wanted to change the name to Metropolitan Area Exchange (MAE) and call it MAE-East.[130] Scott agreed.

Internet lore claims the name anticipated the inevitable creation of a MAE-West[131] in homage to the movie star, but this is unsubstantiated. Scott says he remembers Rick joking about it and Wikipedia makes note of the claim. MAE-West came into being[132] around the end of 1994. These terms are now largely abandoned for the more generic term *Internet Exchange Point* (IXP).[133]

Scott and Rick had developed the concept with a focus on services to deploy on behalf of UUNET. Scott also had to promote the idea internally and make sure it was optimally engineered and installed. This was not an insignificant hurdle to overcome. Initially intended to be housed in one building to reduce costs, the plan had the side benefit of reducing latency between the ISP backbones. The location happened to be an MFS PoP in a parking garage outside of Washington D.C. This was the location of MFS owned equipment, not customer equipment. MFS jealously guarded their PoPs. Allowing customer equipment into this secure space was worrisome.

A typical customer connection would be installed via cables run up a riser to the customer location on another floor, thus the customer did not have equipment in the PoP. The schematic anticipated a building that would not be much used by the Enterprise customers who were the mainstay of MFS. It seems funny today that something so important as the Exchange Point was stuck in a parking lot garage PoP. This illustrates that Internet services simply were not treated as important in that time. Scott says it was chosen specifically because it was not considered important real-estate or a critical location.

Internaut Conference at Morton's

In 1992 and 1993, Rick and Scott would go to Morton's' Restaurant in Tyson's Corner and each have a huge steak while they talked. It was convenient to the MFS Node building where MFS had fiber but also the sales office in the D.C. area. Scott liked to have the 48-ounce Porterhouse and Rick would have that or something smaller like a 28-ounce ribeye steak. They also would order one chocolate soufflé each.

Rick always dressed casually. He wore a blue cotton shirt with no brand or logo and blue jeans. It had a collar like a golf shirt but was just a plain blue collared cotton shirt. It had a faint stain. He liked that shirt and he often wore it.

---

[130] https://en.wikipedia.org/wiki/MAE-East
[131] https://en.wikipedia.org/wiki/MAE-West
[132] http://tinyurl.com/RoutingPaths-MAE-West-111594
[133] https://en.wikipedia.org/wiki/Internet_exchange_point

Rick was ahead of the "*Internet Culture*" as led by Bill Gates and others. This was the beginning of a cultural wave that popularized casual business dress. Professional attire demanded a business suit, especially in the Telecom world. Mark Zuckerberg has led a similar shift toward shorts, tee-shirts and flip-flops in recent years, making "business casual" seem formal by comparison. I think Zuckerberg is still leading that charge with relatively few followers.

Rick would wear his "business casual" to Morton's Steak House, which Scott loved because everyone else was wearing suits. Rick stuck out at Morton's. Scott would take his jacket and tie off and appear relatively casual as well compared to the other patrons. They would sit and plan, looking like outcasts in a sea of suits, all the while inventing the key ideas for the commercialization of the Internet.

They would dine and discuss, explore their ideas and concepts and imagine how the Internet might evolve someday. It was not a given that Rick was right, or that other private IP backbones would accept his ideas, join in and interconnect in this manner. It was not a given that the commercial use of the Internet would catch on with the public, or that business would ever trust the notoriously untrustworthy Internet. These were crazy ideas. In many ways, they still are.

Why would this one little private IP backbone operator and a startup metro-fiber company with a high-speed data services division be able to win versus the big carriers like AT&T, MCI, Sprint, Southwestern Bell, Bell Atlantic, Nynex and so on? It seemed insane.

Scott considered the Internet project as a hedge against the possible slow adoption rate of using high-speed networks in the Enterprise. He also thought it could become a great way to get a foothold into the Enterprise by leveraging the private IP backbone players' sales force. Rick was more focused on selling to other private IP backbone players than selling to small customers or individuals. He wanted to be the backbone for smaller private IP backbone players and sell connections to the big Enterprise users.

Scott loved the idea because it meant selling Ethernet links to both these types of customers and it meant selling more services including local loop circuits. He also foresaw that we would invent other services like dial-up modem services to all these players as we interconnected to the RBOC CO. At this point (early 1992) MFS Datanet was still nascent and the dial-up strategy was not yet a factor, though Scott's notes on the subject make it clear that others were thinking along similar lines. After all, dial-in users needed numbers to call, why shouldn't MFS provide them? Still, though others may have spotted the business opportunity, its potential scale was not obvious. Few were taking this Internet stuff all that seriously.

UUNET Stakes Out the Route

After they came to know each other, Rick confided that if we did all these things then UUNET would become the default route for the entire Commercial Internet. It could happen almost overnight. He recognized that if he used our national backbone and could convince MFS to sell him colocation space for his routers then he could create a Layer 3 TCP/IP high-speed national backbone using the MFS Layer 2 Ethernet as transport.

This approach would then allow him to approach Regionals across the country and persuade them to connect to his routers in each city. It would allow him to leverage the same infrastructure to sell to both regional private IP backbones (ISPs) and Enterprise customers in every city.

He pointed out that the Regional providers (SURAnet, SURFnet, Northwestnet, Sesquinet and others) would be hosting the content and applications that the end user of the Internet wanted to get to via their new "Commercial Internet" connection. This commercial traffic that would **not traverse the NSF backbone** would soon dwarf the academic traffic that had launched the Internet.

The end users would traverse UUNET's backbone to reach this key content because the UUNET network would be the preferred route chosen by the routing tables. It would work that way because TCP/IP automatically chose the "least cost" route, not a dollar cost but a route metric that used the least network resources.

His backbone would be the fastest and best way to reach the desired content via the routed backbone. This would be a coup for him, and since he did not have the cash to build a network without our help, Scott convinced Al Fenn to create a flexible usage-sensitive price option to bootstrap the service with provision to switch to fixed-price once Rick had the revenues to cover the costs.

Scott loved the idea of helping Rick in jumpstarting the Commercial Internet and supported his plan. Scott did not share directly with Al Fenn Rick's plan to jump-start the Commercial Internet through the combination of MAE-East and the national backbone. No one inside of MFS Datanet understood the Internet and what he was attempting to do. He pitched it as a straight-up customer deal. Al was concerned about oversubscription and did not understand that Rick was selling "best effort" and not QoS. MFS could make money even if the 10 Mbps Ethernet were oversubscribed by UUNET. The physical Ethernet port was effectively rate-limiting, satisfying both UUNET and MFS Datanet's technical and financial requirements.

During this same timeframe, they discussed the IP Encapsulation-based Metro Ethernet closed user group concept in more detail. Scott pointed out that Rick and his fellow service providers could create their own peering rules without any higher power defining those rules. He saw a chance to influence the way the Internet was going to evolve and to drive it himself.

It was exciting to sit at Morton's, eating steaks and developing ideas that could make Rick's company huge. Of course, Scott believed they would make MFS huge as well. Scott says he finally understood the true significance of his concepts in the middle of a big delicious porterhouse while waiting on a chocolate soufflé. He was excited that MFS Datanet had a solution and that MFS was the only CLEC who could or would offer these services to the new private IP backbone companies.

The New Rule: There Are No Rules

Rick proposed his plan and made it clear that it would help him but also would be useful to all the other startup private IP backbones. No one entity would "own the Internet." This was a sharp difference to what the NSFNet Partners in ANS were pushing. They sought to put entities such as MCI in charge. Scott agreed and was excited to support the idea because a small group of startup companies was going to make this Commercial Internet happen with a loose set of rules, and MFS Datanet and MFS Telenet would provide all the infrastructure.

He knew this would create a plethora of new customers for MFS Services. He compared it to the Telephone Interconnect Services companies that started selling PBX phone systems and installing them into businesses back in the late 1970s. They created competition with the big Central Office Systems from the monopoly phone companies called Centrex. This competition could be good for everyone.

There was no one person or entity controlling anything the way they were setting it up. We could just go do it and not ask anyone for permission.

Pointedly, Scott did not have to sell the idea to MFS internally because to them he was selling the new MFS Datanet product, a Layer 2 Ethernet service. Internally at MFS Datanet, few understood what they were creating or the potential upside.

In the other parts of MFS, they thought these customers were not credit worthy and too high risk. Al Fenn even told Scott he was wasting time on them. They had intense discussions about spending more time on the Enterprise customers and wasting less time on the startup ISPs. Scott was going a little rogue, going against his own management to nurture this new category of customers.

Scott explained the Telephone Interconnect example and the early long-distance company's examples. He explained why he felt this could be big and why MFS should support these new startup companies. It didn't convince management. He was on notice, if they did not buy enough or did not pay their bills, or went broke *en mass* it was *his ass* that would be chopped off.

Rick and Scott, and no one else, agreed to set up a Metropolitan Area Ethernet, keeping it low-key at first. He renamed it the Metropolitan Area Exchange and it became MAE-East. Rick defined how these new companies were to interconnect to one another based on the agreements he and Scott put in place. Scott told Al and team it was just a closed user group of special companies sharing one Ethernet segment and paying their own connection fees. This was simple, easy and built into the Metro Ethernet Service concept.

He explained that private IP backbones were being built in the D.C. area or would be built by a handful of startup commercial IP backbone players and he could interconnect them all. Scott avoided the Layer 3 discussions and the peering agreements, leaving that to Rick. Scott facilitated it and gave Rick a strong negotiating card by empowering him to take the basic rules they had created together and solicit agreement among the private IP backbone players.

Rick pointed out that these private IP backbones were evolving in each major city and significant market and it was a wild time. They needed to connect to someone who has peered at MAE-East and he planned to be that someone. His company, UUNET/AlterNet was well positioned.

They all needed colocation space for their routers in multiple cities, they needed bandwidth to interconnect their cities, they needed local loops and dial-up circuits to create modem banks. They also needed private local loops to connect their Enterprise customers to the Internet. The field was wide open and no one had a plan. They made one up.

Rick and Scott agreed conceptually to these things over multiple meetings, mostly at Morton's Restaurant. Rick wanted a national high-speed backbone which was much faster than the other independent IP backbones. He needed national colocation space to house his routers in each city. MFS did not have a national colocation product and he needed local loop services to connect end user customers to his backbone. MFS wanted to be a one-stop shop for these needs but didn't have the right products, yet. Scott told him he would have to work internally to get this accomplished.

Dragging MFS into National Thinking

Scott was worried because the MFS Local players were not thinking on a national basis. There was no concept of a colocation play, much less a national colocation deal bundled with local loops, the backbone and so on. There was MFS Datanet revenue for the backbone and the MAE-East Ethernet connections, but the colocation and local loop circuits would go to the MFS Telecom city sales people who did not get credit for sales outside of their cities.

This presented challenges. Since Rick wanted to be the backbone for numerous emerging private IP backbones, he liked the flexibility selling Native Ethernet, fractional Ethernet or plain T1 circuits to connect to his backbone. He also wanted to interconnect with other private IP backbones and with the regional backbones, to sidestep the NSF backbone and its *Acceptable Uses Policy* (AUP), and legally carry commercial content from them. The name AlterNet might have held a clue.

Once Scott understood Rick's UUNET/AlterNet network and commercial needs, it was necessary to devise a straightforward plan that helped UUNET and MFS build their businesses together. The more he grew his backbone and sold connections to new customers in each city, the more orders MFS received. Since he was focused on "High-Speed Networking Applications" Scott saw him as an important way for the MFS Datanet and MFS local loop fiber network to be leveraged and interconnect players to whom MFS was not otherwise visible. Also, a connection into Rick's backbone used a different local loop than the connection into the long-distance carrier network, letting MFS Telecom sell additional loops. It was a win, win, win scenario.

Calculating the Risk

These potential customers, however, were considered risky customers who could go under without warning and not pay their bills leaving vendors such as MFS on the hook for costs and lost revenue. It was clearly a gamble and Scott faced heavy criticism as the company saw more and more shaky startups buying services and piling more and more credit risk on the company's books. New private IP backbone players were just starting to come of age. Scott had the vision and innovative spirit, and believed they were worth cultivating.

Who would survive in this an evolving industry was unclear, particularly UUNET who was underfunded at that time. Scott encouraged the sales force to find more and to sell to all of them in hopes that a shakeout would leave the industry standing. Not ALL companies would fail.

 Scott believed MFS Datanet could sell a great many high-speed data lines to these private IP backbone providers and thought MFS could sell dedicated private lines to connect their customers as the industry evolved. The *High-Speed Data Networking Plan*[134] from July of 1991, makes it clear that selling "off the shelf" local loop circuits complemented having the high-speed data services, creating opportunities to sell more to existing customers while acquiring new ones.

Serious debate continued among industry players about what might happen over the next few years. There was no specific industry accepted plan as to how the Internet would become commercial for the public once the NSF turned off the government funded backbone. It was not clear what was going to happen to the research network. This meant there would be no backbone for any commercial traffic unless someone built it.

Blocking ANS

By mid-1992 it was clear a demand for commercial traffic was growing without funding committed for a commercial private IP backbone to support their traffic. It could not be parked on the NSF backbone with its restricted use policy. From the viewpoint of Rick and other private IP backbone players, it was obvious that there was a drastic need to build national backbones and interconnect them across the US. Rick's ideas seemed to be the smarter and more reasonable. The more players he polled about the Internet, the more these ideas seemed viable.

Many in the Internet community believed that when IBM teamed with Merit and MCI to form a partnership in a new company, the resulting contract would become the commercial backbone. They would own the Internet and dictate usage just as the NSF had — a daunting prospect.

As explained earlier the Merit/IBM/MCI organization formed in September 1990, was called *Advanced Network and Services* and operated the Internet NOC in Ann Arbor, Michigan as part of the venture. Guy Almes, who had been part of the Rice University focus groups in Houston in 1991, went to work for ANS. He invited Scott to see this Network Operations Center (NOC) of the non-commercial Internet in 1992. Scott was familiar with the MFS NOC where they managed the fiber optic multiplexers. He knew this Internet NOC router backbone rode on top of the circuits the MFS NOC managed so it was exciting for him to see the routing layer management system. They could become a customer as well but ANS was tied to MCI so probably not much of a customer. Scott knew about Rick's ideas to end-run all the big boys with a small group of players using MAE-East so he relaxed and watched with interest.

---

[134] http://tinyurl.com/RoutingPaths-1991-Plan

Rick and the other private players such as PSINet did not like ANS, which they considered a threat to the openness of the Internet. There was a joke about ANS whether Rick coined it or just repeated it is unknown. He said, "The only thing missing from ANS is U." This appeared on a popular T-Shirt at Interop in 1992. It was funny yet we had to wait to see if ANS would become the commercial backbone for the Internet. There were other ideas floating around and widespread concern that a big company would own the Internet and force users to pay tolls to get on and off it. Rick and his start-ups had other ideas.

During 1992 debate raged: who was going to "own" the Internet when the NSF turned off the backbone and the small IP backbones started to carry traffic by interconnecting? Rick understood that MCI, Sprint, AT&T, the RBOCs, IBM and others were vying to "own the Internet" (meaning the backbone) after the NSF turned off their government-funded DS3 backbone.

Rick and the others wanted no one to own it. They wanted to allow startups like UUNET to become players even though they did not have the financial resources that the big guys had. Scott loved that idea. MFS was small as well, and fighting the big RBOCs who had billions in revenues funded by their monopoly positions. They were financed by captive ratepayers and a PUC that guaranteed their return on investments. The concept of a protected monopoly was alive and well after the AT&T breakup, albeit in the local arena.

Scott thought the little guys might change the rules and end run the big money incumbents. MFS was still a startup with about 500 employees. MFS Datanet, an early-stage startup, had about 50 employees. MFS was much like these small independent IP backbones in terms of market share, entrepreneurial spirit, and desire to change the rules in an industry dominated by would-be monopolists.

A New Industry

This new type of carrier was evolving out of nowhere. They lacked money and polished business skills, reminiscent of the "Interconnect Industry" that shopped PBX systems around the phone company in the early 80s. The new evolving players who were installing wire and cable inside buildings were called telephone Interconnect companies. They installed the early PBX switches inside the customer Enterprise and connected them to the local loop telephone monopoly as well as the long-distance companies using switched access services via the PBX. This set the stage for the dedicated local loop services provided via the local fiber networks that would be sold by MFS from 1988 through the 1990s. In fact, this was precisely the market *Chicago Fiber Optics*, the precursor to MFS, had launched in 1986 to serve.

Scott had also watched with interest as the early long-distance carriers came out of nowhere between 1979 and 1984. He had considered entering the business himself around1983-1984, but lacked the capital. These new players were struggling and fighting AT&T's big monopoly, competing for long-distance minutes by installing voice switches and tying them together with T1s. It was always about the minutes; minutes meant money. Their network was not a routed IP backbone but conceptually was very similar.

Scott believed these Commercial IP backbones could be the next wave in the industry. Even if the Commercial Internet movement fizzled, he believed significant demand from early adopters would bestow credibility with the large Enterprise customer market.

This is a point that cannot be overstated: data communications services were expensive. Commercial Enterprise, particularly large corporations, had the natural customers. They had the business need for such services and pockets deep enough to afford them. Unfortunately, they were a slow-moving, risk-averse group, difficult for a fresh startup to woo successfully.

On the other side, the Internet entrepreneurs had little money. Selling them anything was a risk. They could disappear at any moment, leaving a supplier on the hook for great expenses.

What they lacked in money and business acumen they made up in entrepreneurial spirit and willingness to take a calculated risk. Successfully delivering a service to them, even if they ultimately folded, forged credibility in the eyes of the corporate Enterprises MFS also wanted. Scott felt strongly that MFS Datanet should be leading the movement toward independent IP backbones interconnecting. It was a risk with immense potential rewards.

This was a completely new business idea, one that could grow in tandem with private Enterprise high-speed data networks. High-speed Enterprise data was the original concept he had envisioned and sold internally at MFS when he wrote the business plan in 1991.This was the same plan he continued to develop in 1992 with Al Fenn after MFS Datanet was created.

The idea evolved very fast. Scott had to convince MFS to support it and demonstrate that MFS could do so without taking a serious business risk. It leveraged to its advantage the entire existing MFS Telecom metro fiber and MFS Datanet infrastructure. It did take a little tweaking and a few new commercial pricing innovations to illuminate these business opportunities. Thankfully Scott could focus his evangelism on the local organizations and did not have to sell the opportunities to corporate management or the MFS Board.

The Telecom and Computer industries did not respect these small private IP backbone players, who they considered a bunch of geeks with no business knowledge. The traditional telecom data people could not imagine using a network that was non-deterministic, which the protocol TCP/IP clearly is. Yet this whole argument would prove to be a straw man, cast aside with the old TDM North American Hierarchy approach to communications, once the Internet took off.

Scott believed Rick and his counterparts understood TCP/IP and saw how it could be used to create a Commercial Internet not dependent on the NSF or other government funding. MFS could play a key role in enabling this new group of players independent of government funding and despite teetered on the edge of being broke every month. He just had to get Al Fenn and his team to provide pricing and support the orders when the time came.

The Unix Wars

Rick was a true computer/Unix/TCP/IP guru, not just a pony-tailed wannabe like so many who followed him. However, he had a legal battle with AT&T over his special version of Unix (BSD) that was draining him financially. These were the days of the Unix wars, before the Linux Revolution.

The original Unix came out of AT&T in the 1970s, itself a revolutionary "skunkworks" project hatched by a small group of people working informally within the company. It was never intended to see the light of day. In the late 1970s AT&T began licensing Unix to outside companies. Tymnet, while I was there, became an early source code licensee, though I am not aware of any practical results from that.

The original AT&T platform was limited. Licensees created their own derivatives that differed from the AT&T version, as well as each other's, designed to address specific requirements. Some of these derivatives include BSD (UC Berkeley), Xenix (Microsoft), IBM (AIX) and Solaris (Sun Microsystems). Many of these derivatives were sued by a litigious AT&T quick to claim copyright infringement on any derivative works. In a sense, something which grew out of an AT&T "skunkworks" project and which AT&T initially tried to suppress then became the center of legal battles as AT&T tried to assert control over it.

Rick Adams was a founder of BSDi, which developed and sought to commercialize the work done on Unix at UC Berkeley. In late 1991 AT&T sued BSDi, the source of Rick's battle with AT&T. The lawsuit was settled in 1994.

Eventually, AT&T divested itself of Unix intellectual property, and the Unix trademark fell to *The Open Group*[135] which champions the *Single UNIX Specification* (SUS).

In addition to variations[136] on the original AT&T platform, clones arose that sought to free Unix-like Operating Systems from the tight rein of the SUS certification. The most successful of those clones is Linux, which itself exists in many variations. SELINUX is a security-enhanced variant used by many serious Enterprise vendors today, notably Red Hat Linux. Though Linux, especially SELINUX, uses none of the original AT&T intellectual property, it does carry forward the counterculture "skunkworks" original in form and function. SELINUX has become a powerful OS, and platform of choice for many uses, including Reprivata and the Community of Trust platform.

---

[135] http://www.opengroup.org/
[136] https://fossbytes.com/difference-linux-bsd-open-source/

MAE IS BORN

Most of these private IP backbones were forced to sell dial-up connections to their customers. Scott showed Al how we could sell UUNET customers reliable and inexpensive 56 kbps, fractional T1, full T1 (1.5 Mbps), high-speed 10 Mbps Ethernet connections and even DS3 (45 Mbps) local loops to his backbone via MFS. This allowed dedicated connection users to connect to the UUNET 10 Mbps backbone.

Scott alone understood the value of MFS selling both MFS Telecom local loops and MFS Datanet Ethernet LAN connections of different speeds and technology. He alone had experience with both kinds of service. It was obvious to him, though not to others. When Scott explained it to Rick, he also understood it.

 The original hand-drawn diagram[137] — sent via fax — includes pricing, monthly recurring costs per location and the network architecture. Note that Scott is telling Al and team to give Rick Adams the flexibility to engineer his backbone using these costs based on Ricks best guess of traffic patterns on that backbone, using the pricing provided in relation to his estimated traffic patterns. This is a combination of business, technical and engineering requirements bundled into a yellow tablet drawing on a copy of an internal drawing of the proposed MFS Datanet backbone sent via a FAX with the actual time and date stamp. This was before MFS had email or Freelance or PowerPoint documents.

 The order[138] for the *Network Access Point* (NAP) later became called MAE-East was logged September 29, 1992. The order was for an Ethernet, not a circuit. The sales representative who worked hard to get the order was Kathleen Davis. She was responsible for booking this order as the MFS sales representative even though Scott negotiated Al Fenn's approval.

The order is really an order form used by MFS Telecom to order a circuit like a DS0, DS1 or DS3. It had recently been changed to include Ethernet as a type of service. The date — September 29, 1992 — is when the sales rep signed it. MFS Datanet did not even finish the business plan until mid-1992. The form shows location 1 and location 2, which do not always apply in Ethernet since it is logically a shared bus between multiple locations, not point-to-point like a circuit. The idea of getting customers to think in terms of a service that was not point-to-point was just one of the early challenges.

Rick liked that fact that he made the rules so only those entities that he approved could join the MAE. All the early MAE players like this because they were in control of the business rules at the TCP/IP layer, not some big "carrier."

Once this order was placed, the oversubscription precedent via MAE-East was set. Unlimited oversubscription endorses the idea that a connection confers the right to use as much as desired. It also means the carrier need not add bandwidth as connections are added. This was the beginning of the "all you can eat" Internet mindset, both for the end user and the ISP. There's no criteria for latency, jitter, packets delivered or other specification of a modern-day Service Level Agreement.

---

[137] http://tinyurl.com/RoutingPaths-MAE
[138] http://tinyurl.com/RoutingPaths-FirstOrder

Scott explained to Rick his worry that the concept of selling customer connections whereby massive oversubscription was assumed set a bad precedent. Scott respected Rick's technical expertise and entrepreneurial spirit but in the spirit of collaborative invention he pointed out Rick was following what Enterprise network customers would think of as poor networking rules and architecture. The design was directly counter to all of Scott's Enterprise networking experience. The Internet was intentionally being designed to be slow and oversubscribed, so badly as to render it in a constant state of near collapse.

Rick explained that the Internet was defined as "best effort." No one expected the Internet to deliver Enterprise-level Quality of Service (QoS). The Internet, he insisted, was all about basic connectivity, not low latency, high throughput or error rate. Rick pointed out that TCP/IP was a connectionless, non-deterministic protocol, so why expect any notion of guaranteed throughput or QoS? There was already the low expectation of "best effort" delivery of packets so why improve on it? He said the user could create as much load as they desired and that flexibility compensated for the unpredictable quality and throughput of an Internet connection.

This established the Internet as an inherently unreliable and insecure network, a perception that still dominates. Even today, heavy congestion often occurs, and the carriers move slowly to augment bandwidth, delaying any expenditure until customer complaints reach a threshold where they cannot be ignored. This is worsened by the billions of mobile devices hitting these backbones and the cell phone network may contribute to the delay and unreliable nature of the connection.

We also wanted to sell UUNET a national backbone, developing that at the same time[139] as the MAE-East order. Al came through and did provide pricing for Rick to buy a national backbone as well. The first order for the backbone was Kathleen's order[140] as well since she was the sales representative who worked for MFS out of DC. Scott signed it for MFS Datanet while Dean Rossi signed it for the local loop division of MFS Telecom and Rick Adams signed it for UUNET.

Before we booked this order, Al Fenn had wanted Scott to "pick a winner" between the small private IP backbone providers. Scott argued that he did not want to pick one since the market would pick them. When Al asked why he wasted time on Rick, who was perpetually going broke, he cited the early telephone Interconnect players and the early long-distance players, pointing out that if we launched them on our infrastructure, they could become huge customers. We could use successful service delivery to these early adopters in order to bootstrap more risk-averse Enterprise customers into giving us a spin.

When Al pointed out UUNET was teetering on the edge of bankruptcy and might not pay their bills, Scott suggested we could turn off their services if they defaulted. Without us, their network did not exist. In the event of a financial crisis, we were the ones they would pay ahead of anyone else. Scott also pointed out that we would not put all our eggs in UUNET's basket. We would sell to, and bill separately, all the participants in the MAE, and later the backbone. If one went under, others might remain afloat.

---

[139] http://tinyurl.com/RoutingPaths-Backbone
[140] http://tinyurl.com/RoutingPaths-BackboneOrder

Al was upset with the whole plan but agreed reluctantly. Al may have felt MFS was virtually betting the company on UUNET's success but Scott thought of them as merely one bet of many. MFS Datanet had the Enterprise customers who were still the focus of most the sales efforts and now it also had this new start-up entity known as private IP backbones, could evolve into something big.

Many leaders in Al's position would not have listened to their VP of Sales. They would have seen these "high credit-risk players with no money and few customers" as a complete waste of time. Although nervous about the risk, Al and his team backed Scott on this issue although he did make him justify at every step that the risk was worth it.

It was a long shot through 1991 and into 1992. Scott rarely talked about these efforts to MFS management since they were focused on the Enterprise. Far from a sure thing, in many ways it was the working definition of "bleeding edge."

John Griebling was new to MFS Datanet and pulled yeoman's duty helping Ken Holcomb and others inside Datanet, a monumental deal at the time as MFS Datanet had no national backbone customers yet and few customers overall. Scott and John were trying to determine pricing profitable for MFS Datanet yet affordable for UUNET. Scott believed UUNET could jump out as the leader of the Commercial Internet with MFS Datanet providing a national backbone and colocation for routers with low-cost fiber-optic local loops to connect to the regionals, other ISPs, and large Enterprise customers. Scott says he could not have convinced Al and Ken Holcomb and team without John Griebling's skill and perseverance in communicating all the technical and network issues internally to the MFS Datanet leadership in San Jose.

Outrageous Oversubscription

Rick Adams wanted a national Ethernet backbone and wanted to install routers all over the USA on our backbone in the MFS cities. Rick needed to build and operate the higher-speed 10 Mbps backbone from MFS Datanet for it to make sense for the other new private IP backbones to interconnect to him across the country.

The de facto position of those early Private IP Backbone (ISPs) pioneers required selling enough customers to create oversubscription of your backbone to make money. One T1 of bandwidth sold 10 times to 10 customers to make money. An ISP would sell 10 dedicated T1 customers but only have one T1 of backbone. Rick knew that if he had a 10 Mbps backbone from us he could sell 100 Mbps of customers in the MFS cities with a mix of speeds from 56 kbps to T1 and up to 10 Mbps to some customers. Reaching oversubscription and even more to get to his 10:1 ratio would take time.

Later, in the 1995 to 1998 date-range, the oversubscription rate became 20 to 1. TCP/IP was designed to handle congestion gracefully, and when oversubscribed would degrade everyone's connectivity in an even-handed manner. Responsiveness may be poor, but things would keep working.

This notion of heavy oversubscription was assumed by Rick and others to be the only way to make a profit while bootstrapping the Commercial Internet they envisioned. This was because they could add revenues without increasing costs if they did not offer any form of service level agreements or Quality of Service as a part of the service delivered. As long as the customer would still buy regardless as to the poor quality, they could do so.

This concept of one connection speed permitting unlimited upload and download capabilities for a flat rate per month, independent of the amount used, was called "all you can eat." It set the precedent to where the user community expected access to all things for one flat rate. It is flawed in the network engineering sense in that it does not consider that some applications or content should cost money above the access fee. It makes sense because these applications create loads that cost more money to support them.

Scott considered it a broken financial model to use massive oversubscription and the QoS would be terrible. Al instructed Scott to discuss the implications of this with me since I oversaw keeping the MFS Datanet backbone from becoming congested. We decided that if the pricing for the 10 Mbps backbone sold to Rick covered the cost when full and it took a while to get there, then if more bandwidth was needed for the Enterprise customers not on the UUNET backbone, there would be sufficient revenue to install additional bandwidth and not congest the MFS Datanet backbone. This allowed us to manage the conflicting approaches profitably and let Rick have what he wanted. Few people know or remember this internal decision-making process wherein we allowed UUNET to have a national backbone at a low startup cost he could afford but still build revenues with usage.

Entrepreneurial Spirit

Rick did not like sales people and, in general, was not especially social. He would be the first to tell you that. He only liked people who could contribute, help figure things out and make them happen. Stan Hanks had told him how Scott conceived the idea to build a fiber network in 1986 and had obtained a franchise from the City of Houston in 1987, and how it took until 1989 to make a deal with Kiewit. It seemed that he believed Scott was not just a classic salesman working for MFS, but someone much more entrepreneurial like him.

Scott had explained how hard he had fought to get MFS to create the Metro High-Speed LAN services and that evolved into MFS Datanet. I think he knew Scott understood how hard it was to build a business. It took a while, but it seems he started to think of him as more of an entrepreneur than a VP of Sales.

Both the Enterprise private backbone networks and the private IP backbone players offered huge networking opportunities. We had to work on both even though the Internet was considered a toy beside the Enterprise at that time. As explained, TCP/IP was considered a "best effort" protocol with no QoS expected or provided, totally unlike Enterprise protocols like Token Ring and FDDI, and the backbone of the Internet made no effort at security, text, email and other data flowed in clear-text format. Private networks used *Time Division Multiplexed* (TDM) circuits, which were dedicated to the Enterprise only, and no one else could use them. You had to build the network one circuit at a time and use a star, ring or full mesh topology with a fixed cost per month for a fixed amount of bandwidth regardless of how you used it.

At that time the Internet was not commercial and no one cared about these shortcomings. This perceived unreliability of the Internet gave value to the higher quality connections we sold to Enterprise customers.

In the end, (by the late 1990s) many buyers would weigh the cost trade-off and migrate toward the low-cost and unreliable Internet. The financial delta was profound and those who valued cost-savings above performance would buy Internet-based services. Even with all the performance and security baggage of the Internet, it was deemed "good enough" by many.

Billing innovations

Rick and MFS Datanet invented the rules for the early MAE with the other private backbone providers (ISPs). We enforced them by not allowing anyone to buy a connection who was not coming in via the authorized group. These unwritten rules perhaps caused problems later, but early on this was critical to attaining the critical mass of participants and kept the incumbent IXC and RBOCs from noticing and crushing us little guys.

Rick liked that it was his "closed user group," as Scott called it, delivered by MFS Datanet. Rick did not want to risk one player bringing the whole thing down if they did not pay their bills nor be in a position of passing through the costs to his IP backbone competitors. Scott suggested each connection be paid for by the company buying the connection to the group.

UUNET would not pay for all the connections and then extract money from their competitors. Instead, Scott created something that did not exist in the MFS circuit notion, multiple payers. Historically, the same party paid for both terminations of a point to point circuit like a T1, even if the circuit connected a customer of our customer. In the traditional Enterprise telecom environment, if a company wanted a dedicated line to a customer, the Enterprise paid the telecom costs and passed that cost on to the end-customer. In the Ethernet model of "any to any" a subscriber could just connect to the bus and reach everyone else on the Ethernet. Scott decided MFS should bill each company directly and each company connecting would pay its own connection cost. This meant that UUNET customers and partners all became our customers as well. This was a radical idea at the time, counter to the general telecom mindset.

The rules about who could join the closed user group stayed in the hands of the members of the MAE, but since each member paid their own bill directly to MFS Datanet, an unpaid bill meant potentially turning off one entity's service and they alone were booted off the Internet, but MAE-East stayed available to the others. This monthly Ethernet bill thus became the most important bill any private IP backbone provider paid in the early days. While everyone did pay, if it had happened, we would have turned off the service. This reduced our risk. We were not solely at the mercy of Rick Adams and UUNET. Our revenue was spread across many players, not in the hands of one. Scott has helped solve a problem for Rick and the early adopters but it also solved a problem for MFS.

This billing concept turned out to be very significant in the adoption rate of the service. A point-to-point circuit has two ends: there must be a single buyer of both ends, but an Ethernet Bus had multiple connections communicating via the Ethernet protocol. The individual companies connecting could pay the connection fee for their own end, and know their traffic was not point-to-point but transported on the Ethernet and seen by all players.

Since Ethernet is Layer 2, it was a natural extension of LAN behavior. We created a commercial/business rule and charged each company a connection charge and saved them from building a prohibitively expensive full-mesh that did not scale. Al understood it and approved it but most people never understood the significance of this approach. This was classically innovative "outside the box" thinking that entrepreneurs are known for and part of why all the early ISPs liked it.

MAE GROWS UP

 By 1994, MAE-East had grown amazingly and was struggling under the workload. The 10 Mbps Ethernet segments were congested and overloaded. Here is a memo from Betty Jo Chang on April 28th showing the concern over the load. Scott was pushing to upgrade the 10 Mbps Ethernet connection to a 100 Mbps FDDI connection.

MAE-East now carried over 90% of all Internet traffic. The Commercial Internet was alive, and rapidly becoming a victim of its own success. It is amazing how small it was in the beginning and how fast it grew from 1992 when we turned up the first services.

The first iteration used the FiberMux Magnum product, which provided an Ethernet segment of 10 Mbps total shared TDM bandwidth. When that filled up, we built a star function of 10 Mbps segments. We discovered that Magnum did not handle wire speed Ethernet loads reliably and had problems with locking up and restarting and other bad behaviors. The testing and validation of the system and upgrades was a major part of my lab activity.

 Later this was fixed but we discovered it since we were the first commercial high-speed hub for shared traffic either Enterprise or public. We were the top aggregate average load and highest peak load location on the Internet. The MAE was the interchange where traffic moved between ISP backbones. Any time the source and destination were on different providers, which was most of the time, the traffic transited this hub.[141] Since no Enterprise on its own had this much load we were de facto the highest speed traffic point for any data networking environment in the world for many years.

We used commercial equipment to internetwork. MFS Datanet was pushing the bleeding-edge limits on technology. When Cisco introduced their 10 Mbps Switched Ethernet[142] product in 1994, called a Catalyst Switch[143] we installed one as shown in this[144] drawing.

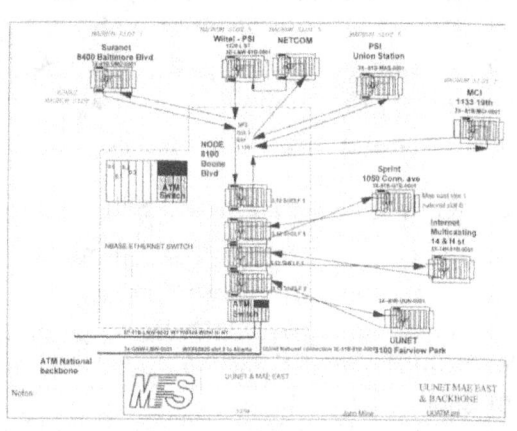

UUNET MAE-East – From website

---

[141] http://tinyurl.com/RoutingPaths-FiberMux-Backbone
[142] https://en.wikipedia.org/wiki/Network_switch
[143] https://en.wikipedia.org/wiki/Cisco_Catalyst
[144] http://tinyurl.com/RoutingPaths-Catalyst

With the advent of the Catalyst Switch, MAE-East became more than one shared Ethernet segment, rather multiple Ethernets switched between each other at Layer 2. Switching at Layer 2 was a novel concept then, and we often had to explain the difference between switching at Layer 2 and routing at Layer 3. Today virtually all Ethernet hubs are in fact switches now. Then, a hub was essentially a passive repeater, echoing all packets on all ports. This made everything on the hub a giant shared media segment and allowed collisions. Also, the shared medium restricted the data-flow to a half-duplex CSMA/CD environment. A switch looks at the hardware MAC address and sends the packet only to the port where the destination resides.

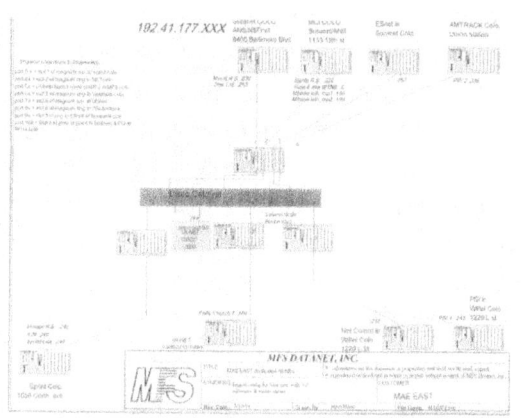

MAE-East Catalyst – From Website

Another difference sometimes is ignored. The advent of switching changes Ethernet from a half-duplex protocol to full-duplex and adds a small amount of buffering. This is a profound change, effectively eliminating collisions and more than doubling available bandwidth on a segment. Now packets can travel back-to-back on the wire without interrupting the flow to receive data in the reverse direction. Further, since each segment only sees the traffic relevant to the destination, the overall bandwidth of a system is increased.

Notice the Cisco Catalyst has a UUNET router hanging off it and then has FiberMux Magnums connected to it which allows 10 Mbps Ethernet to travel over the fiber optic cable on DS3s which are part of the MFS TDM fiber infrastructure to get to other buildings in the D.C. area.

The Enterprise Networking Industry had never encountered applications that could load an Ethernet segment like MAE-East. We often encountered problems that vendors were shocked to learn about. Because we were the first ones to use the technology in such demanding service, we encountered conditions never seen before. The load on the networks we created did not occur in normal Enterprise applications.

NAP vs. MAE

Interestingly, in early MFS Datanet days the UUNET MAE connections showed up on the drawings as NAP, not a MAE, but after a while, MFS reports would show them with the name MAE. Later, the NSF put out a bid to award Network Access Points or NAPs to companies in different cities. They copied the idea of MAE-East and intentionally made the RFP request a Layer 2 service, not a Layer 3 routed service.

It felt awesome to have changed the way the NSF thought about the technology, the role of a peering point and modeled it after MAE-East. This was the idea for a Layer 2 bridged Ethernet backbone based on feedback from the focus groups and shown to Rick Adams on a napkin at a trade show.

MFS Datanet awarded the D.C. NAP

MFS was the MAE-East provider and was also one of the NAP providers. Betty Jo, Bill Euske and I were from the engineering group and John Griebling was our top Sales Engineer at that time. Worked on this proposal while doing our regular jobs was a huge overload, but we would not have been awarded the NAP had all not done a fantastic job. It was developed using a different technological approach and it demanded time to determine how best to create a NAP that was scalable and would not choke from traffic as loads increased.

At a national meeting for MFS, Scott was asked by Al Fenn to pick a winner. Do we keep operating MAE-East or the D.C. NAP? He was standing next to a wall while something social was going on and Al walked over and asked him to choose. He said he had no idea what the market was going to do any more than we knew for sure UUNET was going to become the biggest ISP on the Internet. He said he would just bet on multiple horses. He said that to turn off MAE-East and move everything to the D.C. NAP was not a good idea. The expression "it ain't broke don't fix it" applied in spades. Scott hedged and refused to pick a winner. Al left them both running. This placed the MAE and the NAP in the position of competing for business.

UUNET and MFS Ink a deal for a National Backbone

We all felt MFS had the organizational talent to help make these new services work using new technologies. We were often the first to deploy it to create services for the customers. They were all customer oriented and we used orders from customers to drive development. That was the key. Get orders from customers to generate revenues and then convince the organization to deploy services that met those customer needs. We all believed none of this would have happened inside of AT&T MCI, Sprint or any RBOC.

 Rick had the vision of building his own high-speed backbone to compete with ANS. He did not have National Science Foundation funding. Scott created a special deal to fund his deployment of a national backbone and Al Fenn approved it so it could happen. We finally signed the agreement on December 23, 1992.[145]

The plan envisioned usage sensitive billing to help them grow their business and not pay for the high burst until they needed it. This was totally unique at the time and his backbone grew significantly. You can see the Master Service Agreement defines these unusual terms. This was a totally custom Master Service agreement that defined how individual orders from UUNET would be priced and delivered. Scott had to work with the sales force, operations, and engineering and ultimately get Al Fenn to agree to pricing and approve the terms and conditions in this order. It was not a simple thing and no one had done anything like this before either within MFS or the entire industry.

---

[145] http://tinyurl.com/RoutingPaths-Backbone-Deal

The actual order form for MAE-East uses the term NAP, not MAE. The order has the word "internal" and then it is crossed out and called NAP. So, it was a new concept at the time. Also, the order[146] only shows the 8100 Boone Boulevard location in the notes as one of the other locations.

The sales representative who signed the order was Kathleen Davis. She worked for Dean Rossi who was a City Director for MFS Telecom which sold the circuits. MFS Telecom was the core of what MFS did day to day and generated all the revenues until MFS Datanet launched. Dean worked for Dennis Muse who was a regional manager over the D.C. sales area. He was someone Scott collaborated with often to get UUNET what they needed over time. It was a huge coordination effort among multiple parties inside of MFS to make all this happen.

Scott had to give direct input on how to configure the FiberMux Magnums since he had sold them while at YSA before NCI or MFS and now we were going to be using them with the Enterprise customers and with MAE-East. He marked up an early drawing[147] to show how to interconnect the DS3s in a ring which was an unusual configuration versus point to point circuits at the time.

MFS Telecom had been traditionally what was considered MFS. We had a City Director and sales people in each city whose only job was to sell circuits that connected point A to point B within that city. Scott had been the City Director of MFS Houston from late 1989 until 1991. The Metro High-Speed LAN services was created within MFS; because of that position, Scott was from the inside of MFS. He had the idea to build a metro fiber network and developed the method to deploy and price services before MFS, as Network Communications Inc. When MFS Datanet was formed and Scott moved to the new sister company, MFS Telecom people understood that, and respected this.

Local loop circuit orders were usually generated from an Enterprise customer's office location to their long-distance carriers switch location. This was known as switched bypass and customers chose MFS instead of the local phone company. It was called the local loop or local phone company bypass. Often the Enterprise would build its own private circuit based backbone among all their major sites in all their cities.

We would often sell the local loop from the customer premise to the carrier the Enterprise had chosen to be the carrier between the cities. That was a huge part of the core revenues of MFS. As MFS Datanet VP of Sales Scott understood the way services were ordered, delivered and used by the customers. An example of MFS Telecom services is illustrated in one of the annual reports.[148]

This national UUNET Ethernet backbone order set off tremendous discussion around handling a national account. Local city salespeople like Kathleen must handle components outside their territory. UUNET was a DC-based customer but the bulk of the national order was outside of the D.C. Territory, outside of her territory.

---

[146] http://tinyurl.com/RoutingPaths-UUNET-1st-Order

[147] http://tinyurl.com/RoutingPaths-Magnum

[148] http://tinyurl.com/RoutingPaths-Example

 Dennis Muse from D.C. MFS wrote two[149] memos[150] that explained MFS was going to have to accommodate getting orders from UUNET across the nation for co-location and local-loop MFS Telecom circuits.

Ironically, these memos were faxed between the two groups within MFS. MFS Datanet had its own logical and physical private Ethernet backbone between all cities. We also had our own Microsoft Email system before Exchange was created. It was our efficient method of communicating internally at MFS Datanet.

MFS Telecom, however, had not moved into the digital realm, and used fax machines to communicate between locations. Document generation required typing a memo using word processing software and a PC or an IBM Selectric. Even though Selectrics, discontinued in 1984, were still heavily used by businesses slow to adapt. Once typed, the document then had to be faxed between the locations for input and  approval. Ironic that such memos might be created on a PC lacking email or even faxing capability. Traditional methodology still ruled.

Few people understood how important this was to the MFS Telecom business when this order was placed. It was a major factor for MFS to provide the High-Speed Data Backbone, co-location services, and local loops circuits to connect to Rick's backbone in all the MFS Cities.

Dennis did understand this. He was a big supporter of dealing with these issues, even if it initially created a channel conflict.

An internal memo Scott wrote (and faxed) to others at MFS Telecom about these channel conflict and sales issues to get everyone in both companies to cooperate.[151] He crosses over a lot of internal lines here and was very glad to get co-operation and support from Dennis Muse, Mark Gershien and Ron Beaumont all of MFS Telecom.

Scott had worked for Mark Gershien as a City Director, which bolstered his credibility inside the Telecom group. His pointing out internal sales channel conflict and territory conflict issues in another group inside of MFS was well received and garnered support uncharacteristic of our competition at the RBOCs and the big carriers like AT&T or MCI.

He copied Al Fenn and others at MFS Datanet, including Ken Holcomb who oversaw operations. MFS Telecom people were the powerful people inside of MFS. Their support was crucial.

As a startup inside of a 3-year-old company that was itself a startup, MFS Datanet could not bully its way forward. Scott had to convince them it was in their best interest to enlist their cooperation.

Scott had always made his numbers at MFS Telecom. Even when writing the business plan for High-Speed LAN services in 1991, he made his sales numbers as Houston City Director. He would not have been given the chance to launch these new services otherwise. Datanet would never have evolved. Making net new sales revenue numbers every month was ingrained in our culture at MFS.

---

[149] http://tinyurl.com/RoutingPaths-Co-location
[150] http://tinyurl.com/RoutingPaths-Provisioning
[151] http://tinyurl.com/RoutingPaths-ChannelConflicts

John Hardie joined MFS Datanet in January of 1993 and became an NCC working for Scott. They developed a great working relationship. John told me that one thing that always stuck with him was how humble Scott was. He would never talk about those who worked for him, instead saying, "I don't want people who work for me, I want people who work with me." When I interviewed John for this book, he talked constantly of how much respect he held for Scott.

John stayed with MFS until 1996, when he left to join Equinix[152] where he would replicate the experience of the MAE and create carrier-neutral data centers and Internet exchanges enabling interconnection.

In Jan of 1994 we had an internal memo about the UUNET contract status and what to do next. Steve Luginbill was a regional manager and reported to Scott but John Hardie became the NCC dealing with all these MAE issues including UUNET. Pete Farris was another key player in the region who helped Steve Luginbill and John Griebling get things done. They decided how to deal with the huge explosion of these new type of carriers formerly called a private IP backbone, now called an ISP.

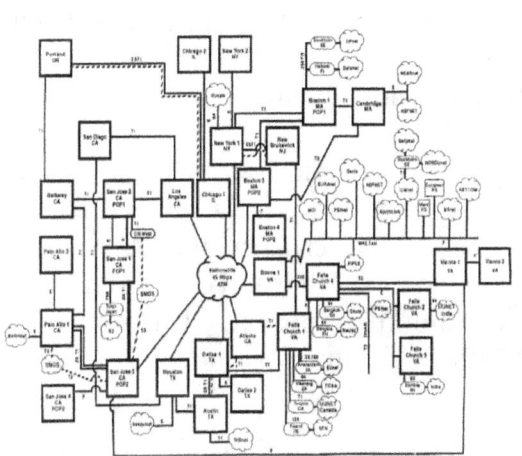

AlterNet • Backbone Configuration 11/3/94

A memo[153] was sent as an email attachment (MFS Datanet implemented Microsoft Mail on the National ATM DS3 backbone with a closed user group Ethernet for internal Datanet use) that explained the original backbone order and issues we had to deal with for UUNET.

A drawing[154] from November 1994, shows how extensive the UUNET network became. The high-speed backbone helped Rick leverage other players like Williams Communications, as well as MFS, and obtain favorable deals for more coverage than MFS alone offered.

This is only two years after that hand-drawn yellow-pad sketch. The many connections are to MAE-East, which also connects Europe and Russia. The cities for the ATM backbone are the cities in the hand-drawn pages and in the internal MFS Datanet page faxed to Al Fenn in November 1992.

UUNET was then in position to make a deal with Microsoft for MSN, which put them on the map. MAE-East is in Boone 1 on the drawing and then there are 2 x E (two Ethernet segments) over to Falls Church another key location for UUNET.

UUNET refers to themselves as AlterNet in the drawing, technically the official name of their network as they christened it in 1990, though they often used UUNET and AlterNet interchangeably. It is an interesting snapshot of the largest ISP on the Internet at that point in time, all transported via the MFS backbone and local loop Ethernet or telecom circuit based services. Alternet routers collocated nationwide in

[152] https://en.wikipedia.org/wiki/Equinix
[153] http://tinyurl.com/RoutingPaths-UUNET-Contract
[154] http://tinyurl.com/RoutingPaths-Alternet-Backbone

MFS facilities connected the regionals so the Commercial Internet provided access to the same content as the NSFNet run Internet of two years earlier.

 MFS included a complete section about MFS Datanet in its annual report for the year of 1994.[155] It names Al Fenn as the CEO and all the services we offered and developed as MFS Datanet are explained. Scott had been involved in all those products and got them approved across the country or they would not have become orders.

It includes a quote from John Sidgmore, who took over as the President of UUNET and Rick Adams became the technology, not business head of UUNET. It explains how MFS Datanet helped them.

MFS Datanet also enabled international LAN-based video conferencing in 1994. The story behind that video conferencing event explains a lot about Networking in 1994, and it used an ISDN connection, which is a story of itself.

ISDN Comes to Sugarland, Texas

Scott installed the first ISDN[156] line in Sugarland Telephone Company history. They filed a tariff with the PUC to install the line, caused over three months' delay, after which the line still didn't quite work. He wanted to call between Sugarland and Houston where there was a second ISDN line connected to the MFS Datanet backbone via a local LAN switch at the MFS Datanet Node. Tommy Waldrop and John Griebling got it working on the SWBT side in Houston.

No one had a bonded 128 kbps ISDN link up and running in 1994. The Sugarland phone company claimed Scott's was the first ISDN line delivered to a home. A T1 solution would cross two regulated local phone companies, crossing the LATA boundaries, at a cost of over $1,500 per month to get a 1.5 Mbps full duplex link to the house. The ISDN solution cost about $100 per month, comparatively cheap since even a 56-kbps circuit would have cost about $600 per month.

A Nationwide Recipe Database

Long before Wi-Fi first appeared in 1997, Scott had an old MS DOS PC plugged into an Ethernet port on his kitchen counter that his wife used for her recipes. John Calhoun spread the word across the country that there were good recipes on the Yeager kitchen PC. One day in the downtown San Francisco node a tech commented to Scott that he liked the family recipe for chocolate cake. Surprised, Scott asked him how he knew about it. The tech explained that all the tech guys in the country checked out the Yeager's recipes. All the tech and operations staff knew Scott had a kitchen PC and two Apple Mac Quadra computers belonging to his sons in addition to his home office PC, visible from anywhere on the national Datanet Ethernet backbone.

That was a great example of how a national network could allow collaboration. It was also a harbinger of the security nightmare the Internet was to become. Even though the Morris Worm[157] had been in the news since 1988, no one thought much about security and privacy in the early networking days.

---

155 http://tinyurl.com/RoutingPaths-MFS1994Report
156 https://en.wikipedia.org/wiki/Integrated_Services_Digital_Network
157 https://en.wikipedia.org/wiki/Morris_worm

A Nationwide Email LAN

Scott would show these bridged computers on the MFS Datanet backbone to customers when explaining its construction, an impressive demonstration tool in the mid-1990s. He would often access his MFS Datanet laptop on the home network as well to track sales activities on a national basis from home, a convenient and powerful resource for tracking 60 sales reps and 40 sales engineers in all major U.S. cities and anywhere from 600 to 1,000 deals active at any given time. Scott could look at all the deals and drill down on the projected revenues and close percentage for every representative in every city as reviewed by the regional sales managers.

Since Al Fenn was CEO and Ken Holcomb was operations, they could pull instantaneous sales reports on the pipeline of deals, the status of the regions and know transparently how much sales activity existed and when it was projected to close. It also gave Scott a view into the 3 big categories of network opportunities in the market that no one else was looking at. Enterprise deals to interconnect multiple locations within on Metro Area, Enterprise deals to interconnect deals across country, Content and Information Service Provider deals across the USA with thousands or local loops into thousands of buildings and Internet Service Provider Deals across the USA. This allowed him to see what was in the pipeline and why deals did or did not happen. One trend was that the cost of the local loop often killed the implementation of the roll out of a new broadband application like video conferencing or streaming video to the desktop.

As a user of the MFS Datanet backbone he was selling, he knew first-hand how well it performed. Also, we were the first company to install a Microsoft email system that worked on a nationwide basis when the MS Mail program was designed to only work over an office LAN. Since we had the High-Speed LAN services, we could build a national Ethernet LAN backbone that functioned as if it was in one building. I am certain that was geographically the largest Microsoft Mail email system ever installed and operated.

Video Conferencing Story

When MFS tech and operations staff sought to demonstrate desktop LAN-based TCP/IP video conferencing over the International ATM LAN backbone between D.C. and London, they purchased two Apple Power PCs with built-in desktop video conferencing. They knew Scott's sons David and Stephen would routinely video chat between their two Mac Quadra's over their home Ethernet and it was effective and easy to use.

Washington D.C. MFS technical people traveled to Houston for a testing session because MFS Datanet expertise originated in Houston when John Calhoun and Tommy Waldrop installed FiberMux gear for customers before there was a Datanet. They were preparing a ground-breaking Trans-Atlantic video conference demonstration, a first-ever video conference between the head of the FCC and his counterpart in England.

Scott was home when he received a call from the tech guys in Houston. It was well-known among technical and operations staff that Scott had an ISDN connection between Sugarland, more than 26 miles away, and the Houston MFS Datanet backbone.

That call was a call for help. Apple computers were unfamiliar in the MFS environment, and the guys were climbing the learning curve. They had spent hours with the video conferencing software but could not get it working. They knew Scott had Macs and hoped he could tell them where they were going awry. Scott went to his son David's room to use his Mac, put them on the speaker phone and invited David, then age 12, to listen in. It required mere moments to set up Apple's peer-to-peer networking over the bridged LAN connection to Houston. Apple computers at the time were much more network-capable than Microsoft Windows machines. Once connected, they shared desktops so each could see what the other saw.

They were explaining what they had tried, with David sitting at the Mac Quadra looking at the remote Macs in downtown Houston. The bridged Ethernet Metro service enabled MAC the non-routable AppleTalk protocol making it as easy to use as a LAN inside a building. He is listening to the adults conjecturing about what to do next. Without asking, David unobtrusively moves the desktop video application on his Quadra Mac onto the desktop of the remote machine, a capability possible due to using the peer to peer protocol. Scott is talking with the techs, and David is making him nervous doing stuff without discussing it. The guys in Houston do not notice and are exploring protocols and/or bridging/routing issues as possible problems. In a minute or two, the transfer completes and all the machines had identical software.

David clicked the application, opened it on his desktop with live local video of himself. Then David inserted himself into the conversation saying, "Click the application on your desktop." They were surprised, asking where that came from? David answered, "I put it there from my computer; I used the network to move it to your computers." This was a remote application installation made simple by using bridged Ethernet over a 26-mile link. Stunning, at the time.

They click on the application. He tells them to pick his name in the application window and when they do, the video conferencing works. A twelve-year-old had used the simplicity of bridged Metro Ethernet to do what he was used to doing in a LAN in his house, without understanding the significance of doing it over a Metro and National Ethernet backbone. He knew how to do it because he routinely worked with both Mac and PC machines on his home Ethernet LAN, and it was no different. The Macs were relatively unfamiliar to the MFS techs but they understood the Metro Ethernet so they grasped that the concept of how a Metro Ethernet Bridged Layer 2 service had made it easy to collaborate, easy to handle non-routable protocols like AppleTalk and it helped them solve a problem in a matter of minutes. It was a perfect demonstration of the value of the MFS Datanet Layer 2 bridged Ethernet service.

More testing followed that initial experiment and the actual demo was carried out with different software. Nonetheless, it demonstrated the value of the MFS Datanet Layer 2 Ethernet service and the guys who built and operated it perceived first-hand the benefit on a real project.

This video conference was also mentioned in the MFS annual report.[158] The video conference reference is on the last page.

---

[158] http://tinyurl.com/RoutingPaths-MFS1994Report

Blue-Alarms

Another anecdote of these early networking days originated in the technical subtleties of DS3 signal encoding and transport. A DS3 was designed for carrying digital voice traffic, not computer data. The early equipment designs took advantage of the characteristics of voice data. Many types of DS3 transport equipment, notably Alcatel DDM 1000 and others based on similar technology, use a type of internal alarm signaling known as "in-band" wherein specific bit patterns are interpreted as error conditions. Specifically, a string of bits of alternating ones and zeroes (e.g. '10101010') or All Ones ('11111111'), depending on specifics, is interpreted as a "Blue Alarm" or AIS[159] Signal.

The random nature of voice data means such a regular pattern of bits would be unlikely in the real world, especially with many (672) voice channels multiplexed into the stream. However, broadband computer data using ATM instead of channelized voice can easily generate this pattern.

This became a problem, one with two solutions. First, ATM interfaces are provided with a feature known as "cell payload scrambling" designed to randomize the data to avoid long strings of specific bit patterns. Second, DS3 transport equipment offers a feature known as "Clear-Channel" intended to eliminate any usage of in-band alarm signaling. Enabling either feature would prevent such problems. However, neither feature was enabled by default.

While initially using and testing ATM equipment on local MFS DS3 trunks, we encountered no issues. MFS used a combination of FiberMux and traditional telecom gear which did not use in-band signaling. However, when we initially launched the Nationwide ATM backbone, we began noticing that the AT&T long-haul DS3 trunks would drop occasionally, flagging a blue alarm. For a while, no one could figure out what was wrong. I quickly cobbled together a test program designed to send pings around a dedicated ATM virtual circuit in order that we might flag alarms before customers could notice, and initiate a hurried reset of the DS3. Steve Feldman contributed a Perl script, and I wrote an MSDOS *Terminate and Stay Resident* (TSR)[160] program, and a quick and dirty cobbled together Band-Aid was applied. With the near instantaneous notification of trouble and a rapid reset, we could recover before anyone else noticed. However, it was Tommy Waldrop who became the real hero, when he spotted the scrambling option and turned it on. Problem solved. The recognition that scrambling solved our mysterious circuit drops revealed that AT&T was not delivering "Clear-Channel" circuits.

The anecdote does not end there. Rather than tell the world what the problem and solution was, we withdrew the AT&T trouble tickets and kept it a secret. We did so because Wiltel and others were following our footsteps in building an ATM backbone and we did not wish to help our competition. For months, afterward, Wiltel was vocal about the unreliability of ATM technology for long-haul transport. We enjoyed the competitive advantage for having the only functional, reliable ATM backbone for a time as other carriers struggled to resolve the issue themselves. The advantage was short-lived but while extant, we gained traction in the marketplace.

---

159 https://en.wikipedia.org/wiki/Alarm_indication_signal

160 https://en.wikipedia.org/wiki/Terminate_and_stay_resident_program

Multicasting and Internet Radio

Rick told EFF Pioneer and "Internet Radio" champion Carl Malamud[161] about us. Carl was an early Internet pioneer and a true Internet guy. He was not rich, wasn't in it for money and was only interested in creating useful ways of using the Internet. His Internet Radio service in 1993 was the first commercial streaming application on the Internet.

Carl was also an unrecognized pioneer in another way. As far as I know, he was the first to serve advertisements to his subscribers, and MFS Datanet became the first Internet advertiser. Although it happened more-or-less by accident, it represents an important milestone in Internet history. It would have required uncommon prescience to see it as a harbinger of the Ad Fraud cybercrime wave to come.

Scott convinced Al to donate a 10 Mbps Ethernet connection to the MBONE (Multicast Backbone) at the National Press Club Building in Washington D.C. The first streaming audio over the Internet originated on our metro Ethernet from the National Press building to MAE-East and then out to the Internet via all the ISPs. Later, we did the first streaming video onto the Internet via Carl and the MBONE. Not very many people knew this would not have worked for Carl if MFS Datanet had not donated the Ethernet into MAE-East which allowed his traffic to hit all the commercial ISPs in one hop via his router.

 In return, Carl put MFS Datanet's name in front of everyone who went to his site, initiating web advertising. He created access to the Patent office and we carried that traffic. He created access to public stock info from the SEC (EDGAR Database). He generated a lot of traffic for stock market company data in Security and Exchange Commission documents or to access the Multicast IP Radio service he created. All that traffic went out over our local Ethernet service and then to the MAE-East hub. This was in place from 1993 to 1997. The MBONE generated significant traffic and came under scrutiny for adding to the congestion.[162]

Humble beginnings

Carl wrote a book about the Internet and included the donations by MFS Datanet as a key to enabling the many innovations he developed in that small room at the National Press Club building. It was little more than a closet, stuffed with Sun Servers and routers. MAE-East was in a PoP in a parking garage.

Many of the first famous sites on the Internet were humble locations, which often surprised visitors. John Hardie relates a tale of a group visiting from Japan, appearing unannounced, asking for a tour of the MAE. When shown the modest assemblage of components in the humble location, they were crestfallen. Our baby had failed to measure up to their imagination. This scenario would repeat at various times when people would show up unannounced asking to see this major nexus of Internet traffic. A group from Germany shared the experience of the Japanese tourists. Peter Lothberg, who visited from Sprint, was similarly disappointed.

---

[161] https://en.wikipedia.org/wiki/Carl_Malamud
[162] http://tinyurl.com/RoutingPaths-MBONEMemo

Backbones collapse with traffic

Once UUNET and PSINet grew to be large ISPs they saw that their backbones were overwhelmed with traffic from the smaller players connecting to them at the MAE. They were carrying the bulk of the traffic so they switched from a policy of encouraging smaller players at the MAE to limiting who could attach. The newer, smaller ISPs of late 1994 and early 1995 were now finding it difficult to peer unless they too had a backbone.

Thus, an aggregator of ISPs was needed.

UUNET and PSI had figured out that they wanted direct connections sold to the business customer. They didn't want the dial-up customer or the new ISP, so they did not like selling to small ISPs. They were business people who had to take care of their business. However, this created an opportunity for someone else to keep the Internet as open as possible. He had learned some things from Rick and understood the importance of MAE-East now that it was the center of the Commercial Internet and everyone wanted to connect directly to MAE-East.

There were several significant email exchanges concerning routing paths, of which this is a sample.[163]

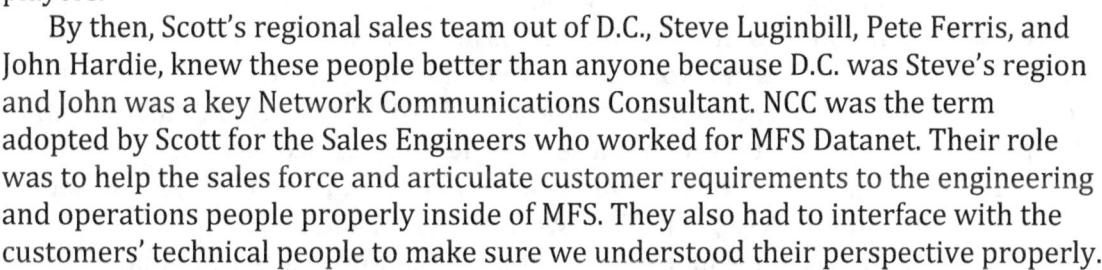

By May of 1994, the traffic on MAE-East was so extensive that MFS installed a Cisco Catalyst Ethernet switch in to augment the bandwidth.

Now they wanted more and for the same price. We had a meeting[164] in the D.C. MFS Telecom Office with all the MAE players to discuss what should happen next. Scott was there, someone from UUNET was there, PSINet founder Martin Schoffstall and his partner were there, as were many other players.

By then, Scott's regional sales team out of D.C., Steve Luginbill, Pete Ferris, and John Hardie, knew these people better than anyone because D.C. was Steve's region and John was a key Network Communications Consultant. NCC was the term adopted by Scott for the Sales Engineers who worked for MFS Datanet. Their role was to help the sales force and articulate customer requirements to the engineering and operations people properly inside of MFS. They also had to interface with the customers' technical people to make sure we understood their perspective properly.

Many Internet players became angry with MFS because the MAE was not performing to their expectations. Since there were no specifications and they just wanted everything to work this created a natural conflict. They were pushing things much harder than anyone ever imagined possible. The FiberMux Magnums were crashing when their segments were flooded with traffic. The segments were running at a utilization of 90 percent and more continuous workload from the MAE routers.

Ethernet was never envisioned to serve that level of workload. Anything above 30 percent continuous and 70 percent peak load was considered maximum when the concept was designed. This was an Ethernet limitation, not an actual MFS Datanet problem. However, perception is reality so we had to deal with it. Ethernet has evolved far beyond the original design.

---

[163] http://tinyurl.com/RoutingPaths-UUCP
[164] http://tinyurl.com/RoutingPaths-UUNET-Opportunies

Steve and John Hardie recruited Scott to come and speak to the ISPs since he had invented MAE-East with Rick Adams and could speak with authority from the perspective of the original intent. The ISPs grumbled about the congestion issues for a while in the MFS Conference room.

Scott reminded them the MAE had originated as a shared 10 Mbps Ethernet segment. We had upgraded to Ethernet switching using the Cisco Catalyst Switch to improve performance at no additional cost to them. Now we simply needed more bandwidth because there were so many players and the traffic per player was escalating far beyond expectations.

They were getting mad at MFS for non-performance when in reality, they had overfilled the Ethernet segments. Scott elaborated that they needed to use good LAN rules to manage the traffic on the Ethernet segments and we had put the Ethernet switch in for free.

Scott explained "getting mad at MFS was like getting mad at a T1 because it was full." He pointed out that MFS needed to fix the problem by putting more bandwidth in the switching fabric, and that it was not reasonable for MFS to do this at no additional cost to them. Mollified, they accepted the logic.

ATM is REJECTED

They discussed the NAP architecture we had developed to propose to the NSF. The NAP had not been awarded nor had the next backbone for the NSF. It was still a bit of a pipe-dream then. Our design was very robust and has been the basis for numerous commercial metropolitan designs for large Enterprise customers. It used LAN (Ethernet) adaptation to ATM over OC3 TDM Circuits on Fiber. Therefore, we would have an immensely scalable network.

We had huge amounts of metropolitan DS3s and OC3's, with OC12 trunks available in our metro fiber network. We were far ahead of any of the IXC carriers and of course way ahead of Bell Atlantic, then the RBOC in that region. However, the ISP players were opposed to ATM because considered it a Telco technology seen as inefficient for TCP/IP. They were IP purists and did not want their TCP/IP packets chopped up and put into ATM cells. They were anti-ATM and balked.

Of course, ATM was much more scalable than other approaches, but that did not matter to this group. John Griebling pointed out the advantages of ATM. John had helped BJ Chang, Bill Euske, and Steve Feldman write the NAP proposal. He was our lead national NCC and later also became the manager of all the NCC's nationwide. He had immense technical understanding combined with excellent communications skills and exceptional ability to articulate concepts and ideas. Scott says John probably influenced the early adopter customers more than any one person.

Though the argument against ATM was bogus, they were unwilling to accept ATM. The TCP/IP guru's thought ATM was a threat to them at that time. This was not something we could get our collective heads around. The technical details under the covers were transparent to the customers. The fact that we were using Ethernet Adaptation onto ATM over Circuits and set up any to any VCI's via ATM was a hidden detail they need not concern themselves with. It was Layer 2 Ethernet and behaved as Layer 2 Ethernet, and how we did it in the middle should not matter. The MAE ISPs got wrapped up in the minutiae and balked at the solution we offered.

Wired Magazine Explains

In 1996, long after this meeting, *Wired Magazine*[165] published an insightful article titled "Netheads vs Bellheads" about the collision between the Internet and Telecom worlds. Clearly the ISPs did not yet have the same respect in the industry that the people who operate the phone networks did and ATM looked like a threat to the TCP/IP world. The way MFS Datanet offered Transparent LAN services was not perceived as a threat. We did not seem like "Bellheads" to them because we were fighting the battle against the "Bellheads" by offering services the telecom world would never have offered then. An internal memo circulated the article.[166]

John Griebling had tested and worked with the key internal MFS Datanet people, including yours truly, to develop the ATM architecture for the NAP. In fact, I had what was probably the most well-equipped ATM testing laboratory in the country at the time.

FDDI Triumphs

ATM, however, was not the only technology I had evaluated in my lab. Since the MFS Datanet team had considered the DEC Giga Router as one potential solution before proposing the ATM architecture to the NSF, John Griebling had a thorough knowledge of both.

The ISPs demanded something LAN-oriented, something TCP/IP friendly. All agreed we needed something faster than switched 10 Mbps Ethernet. We chose the DEC GigaSwitch, a full matrix FDDI non-blocking switch that could handle full speed FDDI segments without choking from traffic. This satisfied all parties as being acceptably "IP friendly" and the best choice. I had tested it and felt it would work nicely. MFS agreed to implement that as the upgrade path for MAE-East. The FDDI upgrade would buy time, but congestion would reappear frequently as an issue. A memo from Steve Feldman on the NANOG mailing list fields one such complaint.[167]

Scott had to convince Al Fenn to let us do this in lieu of using the ATM-over-SONET circuits on fiber approach. He correctly saw it as another technology to keep up with where we could easily use what we already had in our infrastructure. It would cause us support headaches, cost extra money for spare parts and time training the support staff, but the ISPs wanted it their way.

Again, we were customer-oriented and customer-focused. Give the customer what they want if possible, support it and make money. We could do this as a small company instead of fighting over technology approaches the way bigger companies would do with their customers. Despite the headaches, this customer-focused drive ultimately made MFS immensely successful, and the credit for this is primarily due to Scott and his approach to understanding the customer needs and wants.

---

[165] https://www.wired.com/1996/10/atm-3/
[166] http://tinyurl.com/RoutingPaths-Bellhead-Memo
[167] http://www.gossamer-threads.com/lists/nanog/users/8020

MAE-East Becomes a Tight-knit Club

That major decision for MFS Datanet kept the importance of MAE-East in place while allowing MAE-East to grow physically (more ports) and grow in speed (100 Mbps per port and gigabit per second aggregate speeds). This further solidified the importance of being interconnected to a Tier 1 or being a Tier 1 provider even more. It helped the Internet keep growing gracefully without unwanted government or "Bell-shaped" intervention. It was all market driven and rules were defined by the Tier 1 ISPs which was a small club. We followed the Internet culture by letting the early players decide.

I believe this was a key point in time where another company might have tried to control the ISPs before the Internet market fully developed and it could have fragmented or slowed adoption significantly. We got out of the way and did what they wanted at Layer 2, which enabled the ISPs to do their own thing at Layer 3.

The other issue discussed at this meeting was the rules for connecting to MAE-East. Rick had created the MAE and therefore *de facto* defined who could join. He and a few others like PSINet dominated this small group in the early days along with smaller regional operators like Jack Waters of SURAnet, a DC regional network. With their input, MAE-East caught on.

The next issue in the D.C. MAE meeting was how to expand as new players wanted to connect. MFS sales representatives were trying to book orders to MAE-East via sales reps all over the country and via MFS Telecom people. The MAE members wanted to know how MFS was going to handle requests to connect to MAE-East going forward. We had a specific MFS sales representative who was trying to place an order to connect to the MAE from their customer at that time. We did not know if the "group" wanted them in the MAE or not.

Scott reiterated it was their closed user group so what did they want MFS to do. It was a special circumstance. This is another example of Scott making sure the MAE was operated by MFS but controlled by the members. This respect for them was important to them and a very unusual business relationship to show respect for by MFS. Scott had to work internally to keep this respect in place. It was their club.

Scott listened to everyone in the meeting and decided on the fly that John Hardie would be the sole contact for all MAE connections from the MFS perspective. This would mean no one within MFS could book an order directly. It had to go through John Hardie. John worked for Scott and Steve and Pete worked for Scott so he put his trusted NCC in charge of all MAE-East orders and the job of liaising with the ISP's. Scott had decided to officially assign John Hardie to the MAE on the MFS side because we needed continuity of who could get connections.

Scott did this because John was the first Washington D.C. NCC after John Griebling and he understood the unique nature of the MAE structure. This meant he was technically oriented but customer focused. We were dealing with the owners and top engineering people in these startup ISPs and we needed to deploy someone who could discuss things from a technical perspective first; then report back to the sales people and sales management to secure internal approval for providing the services requested. It was a team effort and it was not something that could be made up and enforced by any other company than MFS Datanet. This was a special time.

Upon this request from the Tier 1 ISPs at this meeting in the MFS Building in Tyson's Corner, MFS created an official MFS policy prohibiting any sales representative from selling a connection to the MAE without John Hardie being involved. This was the only way that we could ensure consistent handling of the issues. John had already been handling all coordination with UUNET; now this applied to any connection from any ISP that wanted to connect to MAE-East.

All the early MAE-East Tier 1 people liked the idea that a single person in one company (MFS Datanet) could place orders for MAE-East but put them in control. This single person had to interface with a nebulous group inside MAE-East and find out if they would allow a new player to connect. Scott asked the group what was the rule going to be about who should be allowed into the MAE so John Hardie could enforce the rules. One meeting participant said, "If the person did not know who to contact to get approval inside the MAE then they did not have the credentials to belong in the MAE." This was a very unusual approach to defining membership in the Tier 1 club. John Hardie worked with UUNET and other ISPs every day and understood their way of thinking and unique approach to networking which was different than that of a private network.

MAE-East was a special group with its own rules. We committed to this policy and John Hardie became semi-famous in the Internet community for controlling all orders for MAE-East. The MFS Telecom sales representatives, as well as MFS Datanet representatives, could not book an order to connect to MAE-East without his approval, and he would not approve it until he got approval from the undefined group of people behind the scenes.

MAE-PLUS

MAE+[168] was born in early 1995 as a direct result of this meeting since the ISPs rejected ATM. MAE+ used FDDI and the DEC GigaSwitch tied to the Catalyst Ethernet switch for more ports and faster connections.[169]

Scott and his team had to sell the concept within MFS Datanet. Customers began connecting via FDDI. For the first time, Rick Adams was not running the MAE.

A loose group that only Hardie could talk to would decide who got to join going forward. Like the gatekeeper in the Wizard of Oz, "the Wizard will see you now!" He never let a new ISP talk to anyone directly. They would send an email to Hardie who would just run interference for the small ISP group that decided who could peer at the MAE. If the MAE decided to accept the new ISP they would initiate contact.

Meanwhile, the NSF was funding an interconnection initiative of their own. The NAP was coming. Al Fenn became concerned by the time spent on preparing the NAP bid. He saw his people, notably B. J. Chang, Bill Euske and others such as Ken Holcomb, spending time attending the NAP requirements meetings. They were our network engineering people dealing with product development of services on the MFS Datanet backbone as well as running the existing network. Their time was valuable and limited; Al was right to watch what they worked on.

Al asked Scott for a meeting wherein he insisted Scott must choose either the MAE or the NAP to spend time on and offer to the market. Scott argued we should let the market decide, it was possible both might continue to exist. The MAE seemed endangered, the NSF seemed likely to put it out of business with the award of the NAP contracts. The early ISPs enjoyed control of MAE-East and they did not like that the NSF might define how peering occurred or what rules they might apply.

Al relented and allowed the market to decide. MAE-East grew and became the center of the Internet universe. By mid-1995 it was carrying 70 percent of all Internet traffic and was the place that commercial ISPs interconnected. The growth of ISPs was so wild that inertia and ease of implementation were key factors.

The NSF awarded MFS Datanet one of the NAP locations. It was a great victory to receive the order and recognition as one of the official *Network Access Points* (NAP). Of course, it used the ATM over SONET fiber architecture we had tried to install in the MAE. Since this was the backbone and local loop approach for delivering transparent LAN services for MFS Datanet, it had credibility in the market despite the resistance we had encountered at MAE-East. Ultimately, the NAP and the ATM approach won out and the FDDI-based MAE shut down in June 2001.

Today peering occurs at Internet Exchange Points (IXP)[170] such as Equinix,[171] a worldwide IXC company. Its founders included Pete Farris, John Hardie and others at MFS who worked on the MAE. It is a successful business worth over $30 Billion, built on providing Layer 2 Ethernet connections via some 145 data centers in 15 countries. The model set in place by MFS still operates, run by many MFS alumni.

---

[168] http://tinyurl.com/RoutingPaths-MAE-East-Plus

[169] http://tinyurl.com/RoutingPaths-MAEIP14

[170] https://en.wikipedia.org/wiki/List_of_Internet_exchange_points

[171] https://en.wikipedia.org/wiki/Equinix

Cry Havoc!

Steve Feldman went to a NANOG meeting around the time we were awarded the alternate NAP by the NSF. He announced MFS would move everyone from the MAE to the NAP. No one knew he was going to do this.

Scott suspected Al Fenn was behind this but we doubt that was the case. Rick Adams at UUNET and Marty from PSINet were angry and suddenly left the meeting to go to a bar and decide what to do. They decided to compete with MFS and brought up Bell Atlantic SMDS as a replacement approach. Scott met with Rick, and he told Scott they would use this alternative technology and carrier. Several emails were distributed about this concept. Steve Luginbill, Pete Farris and John Hardie, who worked for Scott in sales, had to contact all the MAE members individually to save the MAE. They had to retract this statement and repeat that we would let the market decide what to do. It was a dicey situation and could hurt our Internet sales if we failed to recover from it.

Steve on the other hand barely remembers the incident today and agrees the described events pretty much match his recollection. He does not recall that Al Fenn or anyone directing the announcement; it may have been simple enthusiasm for the ATM-based approach which everyone saw as more robust and scalable except the MAE FDDI proponents.[172] By the end of 1996, the FDDI head-of-line blocking behavior of the FDDI was straining the MAE and showing us why the ATM solution made more sense. The customer is NOT always right, but they are always the customer.

As Scott learned years before, sometimes what the customer needs, what the customer wants, and what the customer will pay for are not compatible. MAE-East was a classic example of a customer demanding a sub-optimal solution.

In all fairness, the limitations and weaknesses of FDDI were not well known then, and the superiority of ATM was rightly doubted. The best path forward then was less obvious.

Rick's original strategy of not owning the MAE had to be reiterated and reinforced constantly. Even in 1996, MFS did not own the MAE at Layer 3, but did provide the Layer 2 as a transport function. It was a community of interest with a self-identified set of rules, and we sold them a connection to that "closed user group." MFS did not define the rules on who could connect. This body defined it and the body was not a single person. Scott and his team of John Hardie, Steve Luginbill and Pete Farris were forced to mediate internal conflicts and attempts to change the rules on numerous occasion.

This may be the most unique commercial network connectivity situation that ever existed. Each member paid his own bill but no one Enterprise owned the network — very Internet-culture. B. J. Chang, Steve Feldman, and Bill Euske had a huge positive impact with their style as well. They represented the views of MFS Datanet on many committee meetings. Their low-key demeanor coupled with the hard work of John Hardie, Steve Luginbill, Pete Ferris, John Milne, Brian Roberts and many others was a big factor in establishing MFS Datanet's positive reputation.

---

[172] http://tinyurl.com/RoutingPaths-NAPvsMAE

Net 99 and others Revolt

Memos document the thoughts and ideas that developed about the MAE, the evolution of MAE-West and the creation of companies like Net 99, small aggregators focused on providing services to the smaller network operators.

Net 99 was started by Joseph Stroup. We put Joseph Stroup in business overnight. John Hardie, Steve Luginbill, and Pete Ferris had set up a meeting in the hotel room MFS used for the show at Interop 94, which was May 2-6, 1994 in Las Vegas. This was the morning of the big meeting at the CIX[173] in 1994 where people decided that Rick Adams and others might not be treating the smaller players fairly.

Note that by then UUNET was no longer a small struggling company, but a leader in the Internet community and was threatening to the smaller ISPs. Scott had helped Rick become a major player with the special Ethernet backbone priced a special way, provided the national co-location services and local loop services and they had grown very fast. He did well because of the plan and now it seemed he wanted to hold the other small guys down.

 The exclusionary tactics sparked a revolt. Joseph Stroup intended to change the rules at the CIX[174] in response. His plan was to announce a national backbone, to roll out connections to the small players and break up the Internet power struggle between the hegemony of big Internet players, UUNET and PSINet, versus the small startup ISPs.

We wanted lots of orders from many companies. Steve, Pete and John coordinated this meeting at Interop and set the stage for Scott to help the next wave launch on an even footing with UUNET and made a similar usage sensitive pricing deal with Joseph Stroup.[175]

 Stroup was more of a promoter, not a real Internet guy like Rick Adams, but he was a catalyst for getting the next wave of ISPs started. We wanted to get more orders for all the MFS services, get more ISPs and help them get more customers. It took off like crazy.

MFS was struggling to keep up with the demand. After we booked the orders we had to manage issues internally between MFS Datanet and MFS Telecom to get things installed and turned up.

I am not sure the MFS Telecom people fully appreciated how much the MFS Datanet services drove demand for local loop services. We were team players and were properly motivated by our stock options in MFS CC to help all of MFS not just MFS Datanet.

An April 1995 memo reminded the MFS Datanet sales force that they could take the lessons learned in creating a networked approach to interconnecting communities of interest, such as MAE-East, to sell metro deals in their respective cities. Scott had been trying to instill this concept into the sales force for years. The MAE was now in multiple cities across the country as well as other parts of the world. We had growth problems all along and MFS Telecom was not always fully engaged, we had to fight internal politics and lethargy.

---

[173] https://en.wikipedia.org/wiki/Commercial_Internet_eXchange
[174] http://tinyurl.com/RoutingPaths-Net99
[175] http://tinyurl.com/RoutingPaths-Net99-FirstOrder

Net 99 experienced troubles and we had to deal with MFS Telecom and help fix internal issues.[176] We cared about our customers. That was our focus as long as they paid their bills and did not ask for the unreasonable.

This all grew out of filling a customer need and being flexible and responsive to their business requirements. The MAE concept is more unusual for the billing and business approach than it is technologically. To this day, Scott says he has never had a concept take off so well that he used so little influence to get to a tipping point. He said he was convinced that if we tried to manipulate or influence the outcome we would never have been so successful. Our respect for the core group and the rules of this community of interest that made it grow. I think it was our Layer 2 participation that made it possible to stay out of the sights of the "Flamers" on the Internet and earn the respect of the core Internet community. Staying at this level of transport worked to our advantage. It was no an accident. Scott was asked on numerous occasions if MFS should build a routed backbone to compete with the ISP's and he said no. MFS Datanet was neutral and provided underlying services to all the ISP's. Equinix followed this approach because it worked so well.

Even though the MAE concept is important to the Internet (with MAE-East handling over a half a gigabit per second of traffic on a continuous load basis in August 1995) MFS could maintain the perceived "good guy" status versus the RBOCs or IXCs who wanted to dominate and take over, despite our high-profile position. MFS sold bundled or unbundled, local, backbone and special MAE services to the ISPs. We allowed them to pick and choose what they wanted. These were strategies developed by Datanet, and we were in touch with the market.

Clearly, we did have an impact on the industry but it was facilitated by keeping a low profile and not representing ourselves as a big player. We were making money with these services while the industry was saying no one was making money on their Internet services.

Gordon Cook was an early technology and Internet pioneer in the journalism industry, known for his "Cook Report." He made a terrific comment about MAE-East that was quoted in an internal email.[177]

On Tue, 15 Nov 1994, Gordon Cook wrote: "MAE-East works really well, and its settlement free, open peering policies work to the benefit of everyone connected to it. No exceptions! Each connectee pays their own freight. everybody peers with everybody -there's no incentive to do otherwise!"

Scott wrote to his team: "This little-known success story is something all ex-MFS Datanet people should be proud of making happen."

---

[176] http://tinyurl.com/RoutingPaths-Net99Concerns
[177] http://tinyurl.com/RoutingPaths-CookMemo

MAE-West in Motion

The development of MAE–West was another example of how aggressive the MFS Datanet salespeople and engineering people had to be to help that along. John Hardie did a great job of explaining issues.[178]

 *"The biggest difference between MAE-East and MAE-West is the involvement of the Feds. The Federal Internet Exchange Point – 'FIX-West' – is located today at NASA Ames, and is run by a guy named Milo Medin.[179]*

With the transition of the NSFNET backbone to individual NSPs, Milo's main concern is to ensure that FIX-West has connectivity to the three major NSPs – Sprint, MCI, and ANS. Now all the IXCs have a deal with AMES where they have run their own fiber right into the base. This means Sprint and MCI can get to Ames without using any LEC or CAP. So Sprintlink and MCINet have available to them high-bandwidth connectivity at little or no cost. THIS MEANS MAE-WEST WILL EXIST AT AMES.

This doesn't mean that MAE-West can't also be based out of the MFS colocation facility in San Jose. Milo is going to build a commercial interconnection network much like what we have done with MAE-East+ in DC, namely an FDDI ring where companies come and collocate their router and get a connection into the ring."

Scott has talked about how big a deal it was, the barriers to overcome to get MAE-West going. There were location issues, cost issues, technology issues and politics-driven power plays in the industry as players sought to get it at their location. Ultimately, it was installed in the MFS facility in San Jose at 55 S. Market street where I went to work each day. This fiber mentioned by John Hardie around the San Francisco Bay Area meant it functioned as though in one building even though it was geographically distributed. As the distance was not great, about 15 miles direct shot, it had low latency, a big deal to the Internet community.

---

[178] http://tinyurl.com/RoutingPaths-MAE-West
[179] http://archive.wired.com/science/discoveries/news/1999/01/17425

Selling into the Enterprise Customer Market

While pursuing the vertical market of the Commercial Internet, Scott was also focused on landing Enterprise customers. With MFS Datanet approved and funded, lining up early-adopter Enterprise customers became a priority. He felt Rick Adams with UUNET represented one giant customer even if they were not credit worthy at that stage. It was necessary to obtain pricing from Al and develop the proposals to get Rick his national backbone and get MAE-East launched. Scott had gone somewhat rogue to land the UUNET deal but knew if he did not recruit Enterprise customers for the Metro and National Ethernet products at all speeds then the business could not succeed. Enterprise customers were the bread and butter of MFS and the core of the Datanet business plan. This Internet stuff was not mainstream.

Scott had his hands full trying to sell high-speed data services to Enterprise customers. Al Fenn and others had a background in networking Enterprise mainframes in a secure manner. Traditional packet-switched networks were slow-speed relative to LANs and used *Permanent Virtual Circuits* (PVC), considered more secure and reliable. Not only were LANs much faster, LANs did not use PVCs. The protocol TCP/IP used IP Datagrams, which did not flow in structured PVCs but were "routed" based on topology. The TCP/IP protocol was open and well documented so anyone could decipher the packets and look inside of them.

ATM in the Enterprise

Enterprise customers did not trust TCP/IP, fearing it was intrinsically at risk of failure at delivering the information in a reliable and secure manner. The whole argument of deterministic versus non-deterministic behavior was often raised as a reason to distrust TCP/IP. The perceived random nature of IP routing also raised fears of data security. Businesses considered PVCs more secure.

Enterprises did not believe it was a good idea to conduct business over the Internet and did not consider it for any serious business. Today, the concept seems almost shocking, but at the time business considered "the Internet" a fad that would soon collapse.

One of the Enterprise early adopters Scott was pitching early in 1992 was Bear Stearns. Ken Starkey, the head of private networks at Bear Stearns. Ken had pointed MFS at the *Asynchronous Transfer Mode* (ATM)[180] platform developed by MPR Teltech Ltd., as described. ATM became the perfect solution to the "Datagram vs. PVC" conundrum. ATM routes 53-byte (The official terminology is "octets," not bytes, a bit of OSI-speak to confuse the unwary) "cells" of data along Permanent Virtual Circuits and allows the creation of "Closed User Groups" that share nothing but a common pool of bandwidth. Rick Adams could have his over-subscribed, low-quality high-bandwidth transport, and Enterprise customers could have high-quality LAN services and the two would not interfere with each other. ATM was also considered deterministic, for those who bought into that argument, and there were many.

---

[180] https://en.wikipedia.org/wiki/Asynchronous_Transfer_Mode

Bill Euske had recommended ATM to Al Fenn from the beginnings of MFS Datanet. A search was already underway for a suitable platform when the MPR Teltech platform came to our attention. Bill had been privy to some of the early work on ATM in British Telecom and recognized the potential. ATM was the perfect mechanism for converging traditional telephony and the emerging data needs onto a single network.

I was involved in the MPR selection, testing, and deployment. I traveled to their lab in Burnaby, British Columbia and then carted the early prototypes from the Interop show floor in 1992 to be installed in the MAE.

Ironically, however, today ATM itself is falling out of favor and disappearing in favor of an all TCP/IP network. Though ATM still survives in niche and legacy telecom networks, notably the legacy AT&T Uverse ADSL2+, the handwriting is on the wall. Those "untrustworthy" IP Datagrams have won the networking wars, oversubscription and non-determinacy fears notwithstanding. The "IP friendly" transport triumphed in the end. Ethernet now runs at 100 Gbps, one thousand times FDDI speed, and modern routers interface directly with fiber at those speeds.[181]

Even with ATM, the Enterprise market was difficult. We were unproven new players, with unproven new technology. The MAE experience was synergistic, as success would be noted by potential business customers and would encourage them to give us a try.

Enterprise Customers

While chasing the Internet development in 1992 and 1993, we also were building the infrastructure and seeking early adopters for Enterprise customers of the MFS Datanet services. Scott was kept busy helping the existing MFS sales force across the USA get orders from businesses in the buildings that we had fiber into. This was called the Enterprise market.

Scott referred to the Internet as a vertical market. The explosion into public awareness might or might not have happened. It may have remained a tiny niche or fizzled entirely, success was not assumed. The business model was based on revenues from the core Enterprise customers of MFS the companies who bought our bypass services. These ISPs were startups and did not have money and might never make it per management's thinking so they wanted to see results with the customers in the buildings where we had fiber and get them to interconnect their LANs across town and across the country.

The original reason we developed the Transparent LAN services was for "high-speed data networking" needs. The Focus Groups in Houston showed the large Enterprise customers in the accounting, medical, oil and financial services businesses were very interested in private "closed user group" networks. The Internet was just one of the markets that needed high-speed LAN interconnects.

---

[181] https://en.wikipedia.org/wiki/100_Gigabit_Ethernet

Most of the MFS Datanet business was from Enterprise customers wanting to interconnect locations on a private network. We were telling them they could use Ethernet or Token Ring and plug their routers directly into our secure private backbone to create a private network to interconnect their sites. This concept of buying an Ethernet segment or a Token Ring backbone between buildings was simple, easy and cheap.

One of the first things the new VP of Sales had to do was create a sales support position that he decided to call a *Network Communications Consultant* (NCC) after the name was suggested by Ben Gerenstein. He needed twelve NCCs and had to get them on board to help the sales force sell these new kinds of services.

For example, John Hardie was involved in the UUNET opportunity from the beginning. He started as an NCC supporting Kathleen Davis, Bud Caspari and the other sales people in the Washington D.C. market area. Later, when John Griebling came to work for us also in the D.C. area, he immediately differentiated himself as a key player in helping define services driven by customer requirements. He explained the services to customers while he defined the service requirements to engineering and to the operations group. This enabled us to deliver what was sold and allow operations to install it correctly. We had no real standards on the provisioning of services.

We discovered that the New York City network had more early adopter network opportunities than any other region in the US. There were custom applications being developed where the users were looking for sub-second response time at the screen for their trading employees. MFS Datanet put high-speed connections into the trading data feed sites to carry key trading/transaction data directly to the customer trading engines built by these Enterprise customers. Companies like Bear Sterns, Goldman Sachs or JP Morgan were spending millions developing these analytical engines and they needed to make the information available to the individual traders immediately so they could make decisions ahead of their competition which was a competitive advantage.

The Metro Native 10 Mbps Ethernet service was a way to interconnect the workstations directly to the computation engines and data feeds and get the computational responses to the screens of the traders very fast. In the 1993-1995 timeframe, the response time would have been in the range of 300 to 500 milliseconds including router delays and serialization delays on T1 lines.

This was considered fast, so it was a huge advantage if a trader saw a trend or recommendation from the computation engine and decided to buy or sell millions in less than a second. We knew that the delay for Ethernet to the workstation was much lower than the delay through a router to a DSx, especially a DS1. The first time this was pitched to Ken Starkey at Bear Sterns, his eyes lit up as he understood this Native LAN Ethernet was going to give him a competitive advantage especially since his people were geographically distributed around Manhattan.

This also made it clear that selling connections from an Information Service Provider to the desktop of a trader provided another source of information that either helped them make a trading decision in conjunction with their computation engine or some of the data went directly into the custom engine to create a recommendation or data point used in the decision process.

NBC Desktop News

Our engineering group under Bill Euske was always open to new ideas. We supported Scott's customer-driven solutions, even though he was the V.P. of Sales and not an engineer, nor responsible for product development. It was unusual to have such a highly collaborative environment within an engineering company but especially a service provider company. It was encouraged from the top down.

One high-profile example of this collaboration and how sales drove development was the pursuit of a streaming-media service offering for NBC Desktop News. The sales rep was Brian Hayes, and John Griebling was the NCC who drove the project.

The customer needed a connection into each of their client's locations. Those happened to be the traders for financial institutions who wanted to see the news as it developed, in real-time at their desktop. The non-deterministic nature of TCP/IP over Ethernet was considered an obstacle. If congestion occurred, it slows and retransmits, for which video has little tolerance. It will stop and start, become pixilated or both. Various forces pushed Token Ring, accepted as deterministic.

I had a Token Ring testbed operation in the lab and could demonstrate the behavior of both. The first Ethernet Switching Hubs had appeared around 1990 and were rapidly dispelling the myths around shared Ethernet segments. In the lab, it was clear that either technology could deliver acceptable service if not overloaded.

John created the initial drawings, illustrating delivery of the services to the customer. Token Ring seemed the more acceptable high-speed LAN protocol so we proposed it initially.[182]

John Griebling worked up this first proposal with input from Scott and submitted it to Ken Holcomb who oversaw Operations and pricing. John suggested pricing based on our internal charts as this email indicates.

Scott had extensive experience from his YSA days selling LAN cabling. He knew that the twisted pair cable inside the buildings for LANs could be leveraged by new applications so the customers would not need to rewire the building for a video cable (RG6 Coaxial Cable) to deliver video to the desktop.

The buildings had installed UTP cable for LAN use. UTP cable was installed to every desktop by 1993. Leveraging installed cable was a huge advantage for NBC Desktop News in their sales cycle. Scott explained it would be available at minimal additional installation cost to the Enterprise by leveraging the existing LAN cabling.

One of the side-effects of this project was that I was to become an expert on structured wiring. The original issue of the TIA/EIA-568[183] standards came out in 1991, and 'Revision A' in 1995, followed by 'Revision B' in 2001. The current revision as of 2014, is 'Revision C.' The revision is not to be confused with the termination pinouts, designated T568A and T568B, a confusing similarity of nomenclature.

I attended a special telecom class held in San Francisco, becoming MFS Datanet's expert in the wiring technology. I still have the binders, documents, and tools, and have used the expertise on many occasions since. This concept of leveraging the LAN cabling removed a sales barrier for the content provider.

---

[182] http://tinyurl.com/RoutingPaths-NBC-TR
[183] https://en.wikipedia.org/wiki/TIA/EIA-568

MFS Datanet offered a tremendous advantage to NBC Desktop News if we could deliver video over the metro local loop using the high-speed HLI data services. At the network demarcation point we would hand off the video traffic to the Enterprise LAN and they'd deliver it to the desktop. This concept applied to many applications but was especially obvious for video.

A requirement to pull new cabling inside the building to every floor and every workspace was a non-starter. The issue was holding back the deployment of numerous new applications, especially video to the desktop.

This first version of a pricing quote created internally by John Griebling was the Token Ring version. The NBC engineering people wanted to make certain the video traffic was carried in a quality manner so there would be no jerky motion video or freeze frames. This was an attempt at getting the video stream to the end users with high network *Quality of Service* (QoS), in 1993.

That concept was very new. No one was delivering streaming video over a LAN. It was not thought possible over LANs because video was an analog thing and not digital in the minds of most users. The concept of delivering quality video over a LAN was radical at the time.

Al Fenn considered it a waste of time but he allowed Scott to run with it because it was NBC Desktop News and Scott understood the cabling issue in the buildings. He knew we could change the game with the HLI services, so we took it as a challenge.

Since the MFS Datanet Frame Relay switches were online, we could use Frame Relay in a non-oversubscribed manner, especially in the local loop, and we shifted to Ethernet over Frame Relay, developed with Al's approval, May 21, 1993.

The entire sales team including Cynthia Bissett Germanotta, John Griebling, and Scott all solicited client feedback while crunching the final version. John Griebling produced the final version of the proposal and coordinated internally with the sales organization, Bill Euske and our team.[184] Ken Holcomb was working with Al Fenn to get the final pricing approved while Scott was the voice of the customer and represented the feedback from the sales force. Scott considered it his job to get what the customer wanted approved in those internal negotiations. Griebling modified the proposal and submitted the final version on May 25, 1993, to the VP of Technical Operations at CNBC.[185]

It was a gutsy move to develop and deliver this solution to NBC, who was going to sell it to their customers all over the USA. There was a time in late 1993 when NBC Desktop News was doing a road show to sell their services all over the USA. They came to a Houston hotel with a live product presentation to over 150 potential clients, and we were having some problems internally.

Scott was sitting nervously in the back of the room, knowing our people were scrambling, fixing it real time. They resolved the issue moments before the live demonstration; the demo went beautifully. Streaming media to the desktop of a PC was now a commercial service offering. Streaming video to a PC was dramatic.

---

[184] http://tinyurl.com/RoutingPaths-NBC-Final
[185] http://tinyurl.com/RoutingPaths-NBC-Revised

Bloomberg was a huge customer of MFS Telecom and they sold the Bloomberg Green Screen Terminals for over $1,200/mo. at that time using a 56kbs circuit.

We had NBC Desktop News in production in 1994. We used Ethernet adaptation to Frame Relay for the local loop with Frame Relay over ATM on the long haul between cities. The technology was transparent to the users who simply plugged in an Ethernet cable into a LAN switch. This was an innovative and elegant solution put together by the MFS Datanet team, working to meet the customer's needs.

 That year we won an award for this innovative new service at Interop. The year before we had won the Hot Products award[186] for the LAN services with our High-Speed LAN Interconnect service. Now we won it again[187] for the Streaming Media Services to the desktop for NBC. Winning the Hot Products award in two consecutive years was quite a rewarding experience. Bob Barbour did an outstanding job of making sure the trade press knew of us. MFS as a company and Datanet as a group within that company were constantly innovating and making new things happen. I am proud to have played a role in making it all happen.

 NBC Marketing produced a beautiful color brochure[188] touting the service saying, "NBC Desktop Video chose MFS Datanet's High-speed LAN Interconnect (HLI) Service to deliver their applications. MFS Datanet's fiber optic network uses the most technically advanced and innovative technology to provide up to native LAN speed interconnection seamlessly and easily."

Hoot-n-Holler

One Enterprise customer had purchased an Ethernet backbone from New York to Chicago and then to San Francisco. They were a stock trading company and wanted to have live "hoot-n-holler"[189] capabilities between their locations. This was an old Telephone/Telecom term for a permanently open line that acted like a party line from the early analog days of the phone company. It allowed people to have an always on phone line between all the locations where it existed. It saw usage in many industries, from stock traders to auto parts salvage yards. If you needed something not in stock locally, the attendant would just shout on the "wire" and maybe another vendor miles-away would answer back.

Traders could shout to the person or people on the other end and say buy or sell in a flash without having to pick up a phone and place a call. MFS Telecom sold thousands of these between buildings in downtown Chicago and New York for the financial and trading companies.

---

186 January 1993 Data Communications Magazine "Hot Products" HS LAN Interconnect
http://tinyurl.com/RoutingPaths-HotProd
187 January 1994 Data Communications Magazine "Hot Products" NBC Desktop Video
http://tinyurl.com/RoutingPaths-HP-NBC-DV
188 NBC Color Brochure http://tinyurl.com/RoutingPaths-NBC-DV
189 https://en.wikipedia.org/wiki/Hoot-n-holler

Using traditional telecom technology, it was a decades-old and well-understood product. Bringing it into the computer data communications realm was novel and innovative.

If the customer wanted traditional telecom long-distance lines between cities the proposition became expensive. Keeping a long-distance line "open all the time" could cost hundreds of dollars per month versus about $50 per month for a local "hoot n holler" line in the metro area.

Scott discovered a small software company in Houston that had created a "hoot-n-holler" service between PCs as an application running over the Ethernet. This was Voice over Ethernet, what today we might call VoIP. They initially used it over IPX since it was a Novell LAN but later converted it to TCP/IP as the application protocol. In 1994 this was a precursor to modern *Voice over IP* (VoIP) services. The company was *Modulus Technologies* and the owners/software developers were three brilliant guys named Rex Shelby, David Berberian and their software guru Larry Ciscon.

Scott wanted to replicate this and sell hundreds of Ethernet services loaded with "hoot and holler" point to point or point to multipoint PCs on the LANs in either the Metro Area or on a National Ethernet. The Telecom Operations people and even the MFS Datanet people thought it was a crazy idea. They laughed at the thought of putting voice over an Ethernet service. Scott finally convinced John Calhoun and some others, explaining the differences between a heavily-loaded LAN Ethernet and a lightly-loaded wide-area LAN connection. If bandwidth was not overloaded, there would be no collisions and the QoS would be satisfactory to make the voice work adequately. By then he could point to a real customer who had been doing just that over our national Ethernet for over a year. Further, with the acquisition of Kalpana[190] by Cisco in 1994, full-duplex switched-Ethernet switching was being rapidly adopted which eliminated bandwidth-robbing and latency-inducing collisions. Ethernet behaved in a more deterministic manner.

Scott encouraged sales forces in New York and Chicago to identify potential customers receptive to a demonstration of the concept. We put the system in service on desktop PCs and in each city to show it worked. Cynthia Bissett Germanotta was the first to identify an opportunity in our friends at NBC.

This was the first commercial Voice over IP service, and it began in 1994. There are several internal memos discussing the service and the meetings involved.[191] An email trail between Cynthia and Scott records thoughts about the possibility of a trial. "We have a perfect prospect in NYC for the 'hoot and holler' application and they are willing to be a beta site and work with us ..."

A fax[192] from David Berberian and Rex Shelby to Steve Lee and Scott Yeager on the meetings of October 18, 1994, "*We thought it would be useful to summarize the key points from the MFS Datanet/Modulus Technologies meetings. These notes pertain to the following two recent meetings: Meeting ... in Houston on October 12; Meeting ... in Chicago on October 13 and 14.*"

---

[190] https://en.wikipedia.org/wiki/Kalpana_(company)
[191] http://tinyurl.com/RoutingPaths-Hoot-beta
[192] http://tinyurl.com/RoutingPaths-NBC-dv-Notes

We presented the demo to John Bloomer with NBC, having shown him the first Streaming Media to the Desktop in 1993. A fax[193] from October 23, 1994 provided follow-up confirmation of that meeting. Another fax[194] sent a month later, on November 25, further described the capabilities discussed and next steps.

The "hoot-n-holler" service and Modulus Technologies would also become a core component of Scott's Metered Application Service (MAS) concept, which he first presented to MFS and then later took to Enron Communications Inc.

Going Low via Frame-Relay

We recognized that not all Enterprise customers could afford to buy our full speed HLI service to connect their locations at LAN speed. We needed to include less expensive lower-speed services in our armamentarium to encourage customers to think about the ease of interconnecting distant locations. We could upsell them to faster speeds over time, but offer a lower initial barrier to entry.

We also needed to illustrate that as their requirements evolved, they could scale effortlessly to ever faster speeds. There is a terrible gap between Ethernet LAN-speed and the next lower increment in the telecom hierarchy, the T1 speed. Connecting to the data network at T1 speed was challenging, as there was no good way to get from T1 to DS3 ATM.

If a customer merely wanted to connect two offices in the same city, MFS Telecom could sell them a simple T1 at highly competitive prices. But once they needed to connect nationally, or to connect multiple locations they must transit the ATM network, and that's the tricky part.

The lack of a cost-effective low-speed solution was slowing down the sales cycle. We needed to deliver a less expensive entry-level network solution.

Al's team of engineering and operations people were building a Layer 3 internal network for operations and management. Because of the influence of the Internet customers and the growing popularity of TCP/IP, some were championing an all Layer 3 routed solution for the low-speed aggregation issue. In the minds of some, low-speed aggregation was a natural outgrowth of the routed network we were already building.

While we had need of Layer 3 routing for some applications, including our own internal network, this meant taking our customer space into Layer 3 territory. This would be a new direction from our previous Layer 2 only position. This presented operational challenges, as well as potential confusion in the customer base. Until this point, we had stayed purposefully in the Layer 2 space.

The T1 port density of available routers was too low and the per-port cost too high. In addition, no available routers had a DS3 ATM interface, making connecting the Layer 3 network into the ATM network a challenge.

---

[193] http://tinyurl.com/RoutingPaths-NBC-dv-followup
[194] http://tinyurl.com/RoutingPaths-NBC-dv-needs

In my lab, I was working with several router vendors who were promising DS3 ATM connectivity soon, including NetEdge Systems, Inc.[195] We were investigating possibilities and playing around with early prototypes. Nonetheless, even if we could obtain the necessary hardware with the right physical connections, the port density was simply too low to make the service practical. Besides the hardware costs, we had to account for the management costs. We needed to train operations techs and get our provisioning people and systems up to speed before introducing a routed solution to the customer base. The Layer 3 approach was problematical from several directions.

Tommy Waldrop had performed preliminary investigation on another approach. He suggested Frame Relay as an attractive aggregation technology, as it allowed high port density at low cost per port and allowed us to define the QoS over the local loops into the backbone, all while staying squarely in the Layer 2 space. Frame Relay was at the time used only by the RBOCs as a platform for Data services but was not generally well-regarded. It could aggregate numerous lower-speed customers onto the national ATM backbone efficiently, and it could be cheaply populated with many T1 ports and a DS3 ATM trunk. This was yet another example of Tommy's outstanding focus and dedication.

Cascade Communications,[196] founded in 1990, was a new company that had decided to capture the nascent Frame Relay market. They built an efficient and compact switch that combined Frame Relay and ATM in one chassis, making it easy to aggregate Frame Relay onto our backbone. Tommy Waldrop convinced Scott to meet with its representative Bob O'Neal, presented convincing arguments about the value of frame relay and strong justification to use it in our network.

My boss, Bill Euske, agreed we should look at Cascade. Scott wrote a memo to Al Fenn after the meeting with Bob, defining the issues so Al's team could make informed engineering, operations, and financial decisions. It apparently influenced Al Fenn, as I almost immediately had a shiny new toy in my lab. Frame Relay proved to work well, and we began rolling it out almost before I could sign the test reports.

Cascade was just a startup at that time and had not been selected by any RBOC. It was risky to select a vendor that was unknown. Frame Relay had a terrible reputation in the industry. It wasn't the technology's fault, nor was it the fault of the equipment manufacturers. It was the fault of the operators who permitted massive oversubscription and did not assign someone like me to the job of managing bandwidth in the backbone in relation to usage so it was not congested.

Frames Behaving Badly

The Frame Relay networks deployed by Sprint and Williams Communications were built on T1 backbones, not a DS3 ATM backbone as ours was. Imitating the business model of the Internet operators, they seriously over-subscribed the T1s, resulting in terrible congestion and high latency with unpredictable congestion occurring routinely. Serialization delay due to the T1 vs. our DS3 trunks was a factor, and switch delay through all the switches was high because of the congestion.

---

[195] http://www.bloomberg.com/research/stocks/private/snapshot.asp?privcapId=32009
[196] https://en.wikipedia.org/wiki/Cascade_Communications

Frame Relay switches have relatively large buffers, unlike ATM switches which hardly buffer at all. Under congestion, an ATM switch may discard cells if it must, but it does not delay them. Transit time across an ATM cloud is relatively constant, as cells are either lost or delivered, not delayed.

If a line is congested, ATM will discard the cells and trust the higher-level protocol's recovery mechanism. TCP/IP has a well-designed and efficient recovery mechanism, thanks largely to fine-tuning under fire in MAE-East. TCP was designed for an unreliable network. IP is a datagram protocol designed to either be delivered or discarded. TCP was designed with the assumption that frames will be lost over the unreliable IP protocol. TCP expects an acknowledgment returned for a frame within a certain time, expressed as a function of the line speed. This is called a "sliding window"[197] and failure to receive a timely acknowledgment is called the "window closing." Thus, a DS3 line has a much shorter time before the "window" closes than a T1. This is called the "Bandwidth Delay Product."[198] If the window closes, TCP assumes a frame has been lost and will retransmit to recover the lost frames. It will also slow down, probing slowly as it tries to recover from what seems to be a loss of data.

Frame Relay switches will buffer under congestion, trying never to throw data frames away. This means that TCP/IP may attempt to recover "lost" packets that were not lost at all, and in the process, stuff the buffers with retransmissions, further slowing everything down. The same "lost" packet may be delivered several times after increasing delay. Recovery cannot begin until all the buffers have been emptied, or flushed.

This behavior is called "congestive collapse,"[199] and it increases latency significantly. Adding switches adds even more latency, since each switch adds buffering. Pushing a network into congestive collapse floods already congested lines and buffers with more and more wasted retransmissions until the TCP protocol has slowed to a virtual stop, after which recovery can take a long time. Extensive work fine-tuning the protocol has minimized this behavior.

MAE-East had encountered congestive-collapse because the congestion there pushed TCP in a way it had never encountered before. Rick Adams led an effort to fine-tune TCP to work optimally under heavy congestion. Research performed by Van Jacobsen[200] figured prominently in this work.

Tymnet had also encountered similar experiences in the early days and had tackled it with protocol fine-tuning, but neither case had the buffering and Delay Bandwidth Product of a nationwide Frame Relay network running over DS3. We understood this mechanism and engineered our Frame Relay network to avoid congestion, thus avoiding the bad behaviors Frame Relay was known for in the RBOC networks.

---

[197] https://en.wikipedia.org/wiki/Sliding_window_protocol
[198] https://en.wikipedia.org/wiki/Bandwidth-delay_product
[199] https://en.wikipedia.org/wiki/Network_congestion - Congestive_collapse
[200] https://people.eecs.berkeley.edu/~fox/summaries/networks/cong_avoid.html

Frame Relay in Production

Bill Euske set up the meeting with Cascade, with an open mind, and let me have my way with it in the lab. If he hadn't, we might have gone the router approach, in which case we would not have had an economical way to sell lower speed services. Having a head of engineering be so open-minded was a big deal, not something commonly found in the traditional "Bell-shaped" world. It was a huge advantage to have this leadership from Al and his management team.

From the sales perspective, MFS Datanet needed a low-cost fractional service at T1 speeds and below. It was essential that we reduce the cost of aggregation. We could do this if we could get Ken Holcomb to commit to offering fractional speed (at from 56 kbps to 512 kbps) Ethernet services built on top of Frame Relay.

We would use frame relay to aggregate traffic onto the IXC backbones as well. Al assigned Jay Jonakait to that idea. From it he developed *California Express*. California Express was a tie from our local network to the local RBOC network to reach small operations that were not on our network. It was popular, though its main value was reaching small remote offices of larger customers.

And then came SNA

The concept of using the frame relay to provide services for SNA networks was also something we pursued. MFS bought Cylix Communications Corp. out of Memphis Tennessee and used their technology to reach into the IBM user base.[201] As I recall, the product was mostly sold to trucking companies.

I recall traveling to Memphis in connection with this effort and working with Cylix in their facility. I have forgotten much of this project, but I remember two things from that trip. First, they had enormous datacenter backup generators built around a pair of Allison V12 piston engines,[202] which I found impressive,[203] and second, I managed to visit the original Memphis Belle bomber which was on open display in a park in Memphis.[204] I have always been fascinated by WWII aircraft and was deeply moved to walk up and touch that great warbird.

When this developed, Scott oversaw a focused special sales group that managed this business and grew it using an SNA extension product called SLI.

Selling SNA networks was different than Interconnecting LANs but the installed base of SNA-speaking IBM computers was significant so it made sense to purchase a company with a presence in that space.

---

[201] https://www.thefreelibrary.com/MFS+announces+acquisition+of+Cylix+Communications+Corp.-a015906539

[202] https://en.wikipedia.org/wiki/Allison_V-1710 - Technical_description

[203] https://www.youtube.com/watch?v=XYfj-SIyn1Q

[204] https://en.wikipedia.org/wiki/Memphis_Belle_(aircraft)

MFS DATANET IS ASSIMILATED

In August 1995, management instituted a reorganization of MFS. Datanet was folded into the overall MFS organization and was no longer a separate entity. Several of us found it difficult to deal with this since there were so many good people at MFS Datanet, but MFS Corporate leadership decided we were not generating enough growth to justify the cost and effort. I believe they did not realize how much revenue Datanet had funneled into the Telecom organization.

The services and the network continued to be offered under the new umbrella of MFS Global Network Services.

UUNET Joins the MFS Family

On April 30th, 1996, MFS announced its purchase of UUNET, then the biggest ISP in the world. Already a MFS customer, they had a huge installed customer base. I do not think upper management understood the Datanet backbone was a private Layer 2 backbone that let any protocol run over it even non-routable ones. It was not exclusively an ISP backbone but was underneath the ISPs Layer 2 Ethernet networks at ATM level.

I am certain they did not care or understand that selling private data networks to big Enterprise customers was different than selling a connection to the Internet. By then the Internet was capturing mindshare.

The Internet was a "network of networks" all interconnected. It was highly open to attack from anyone on the Internet from anywhere in the world, although to be fair, no one anticipated the nation-state sponsored attacks we face today. The advantage of global visibility from any source to any user or between any users was and is a huge liability. Anyone on the Internet can also attack any other source or individual if they know what they are doing. This risk was being lost in the "race to the bottom" in pursuit of commodity networking. The Internet, with all its flaws, was winning the culture war.

MFS and UUNET were two different products and businesses, and both were needed and in demand. I think everyone was going crazy by then to be an ISP so buying UUNet made MFS an ISP overnight and seemed to negate the need for MFS Datanet. We all thought they were wrong. Scott was then just a VP of Sales in a subsidiary with no product to sell.

 MFS was public, Jim Crowe and Royce were darlings of Wall Street. The stock was doing well and buying UUNet was a coup in the eyes of the public markets. MFS Datanet was a detail. Scott took it philosophically, but he knew what MFS Datanet had done was special and he told his people they had done a special job making it happen.

The Last Mile, DSL and Buckets of Cash

MFS went public in May of 1993, partially separating itself from PKS and gaining a cash infusion from the public underwriting. Due to the continuing expenditures fueling expansion, MFS had been experiencing continual net operating losses. In 1992 they lost $13.1 million on revenues of $108 million. The 1993 IPO brought in approximately $1 billion from the sale of stock, plus a debt vehicle. PKS retained a little more than half of the MFS shares, the rest being sold to various investors, including the public. PKS had invested something around a half-billion dollars of its own money into the company prior to the IPO. The new money was committed to growth and MFS began aggressively marketing its services, moving into new areas and downscale to find customers beyond the fiber footprint.

One of the areas of growth funded by this new capital was the construction of and later upgrades to the first nationwide Asynchronous Transfer Mode (ATM) network. This aggressive use of new technology enabled transmission of voice, video, and data signals simultaneously on a single backbone. The establishment of nationwide ATM service evidenced the company's ability to offer more sophisticated telecommunication services than other much larger rivals, such as AT&T, MCI, and Sprint.

ATM was a bold move; the technology was barely in existence. An important requirement mandated not only a robust ATM platform, but both Ethernet and Token Ring interfaces (and later FDDI). The platform under development by MPR Teltech of Vancouver seemed to meet the requirements. After the initial review by the executives, I flew to Vancouver, BC in 1992 to evaluate the technology in the vendor's lab. It was almost ready, but the Ethernet interface was a bit fragile and the whole system had a flavor of "engineering prototype." Still, it worked and the decision was made to proceed.

In October 1992, MPR Teltech Ltd. presented their new switches at Interop in Moscone Center in San Francisco. They literally displayed the only working prototype switches in the world at the trade show. Other manufacturers had bare-bones switches, but the MPR switches had usable interfaces for Ethernet and Token Ring, critical features to our plans. At the close of the show, I loaded the prototype switches into the trunk of my car, carted them away from the show, and shipped them to be installed in the first network locations. I was proud to have played a major role that first ATM implementation and wrote some management tools and some creative "glueware" to allow the MFS Datanet Network Operations Center to manage the network and detect outages. There was a fair amount of "bailing wire and shoestring" construction in that first system, but despite that it worked beautifully.

One of those first locations, of course, was MAE-East, and soon we had an ATM Backbone carrying national traffic from all the ISPs joining in with UUNET. The ISPs had rejected ATM for their local interconnections, used it for the long-haul national backbone because there was no other option.

Expansion of the Internet came rapidly, and we had all the pieces. We had the national backbone and the only MAE on the planet. We had modem banks for dial-up modem-based Internet users to dial into without needing to call long-distance, and we had local LAN-speed connections for large Internet users. When the Frame Relay network came online in 1994, we could offer lower-speed, and thus cheaper connections to companies that wanted a dedicated connection but could not afford the high-speed LAN style connections.

Even a fractional T1 using Frame Relay was still an expensive prospect for many early Commercial Internet adopters. We needed something even cheaper still. The Telecommunications Act of 1996 was coming together rapidly, and we were salivating at the prospect of gaining access to the ILEC local loops. The Interconnection Agreements that Andy Lipman and team were negotiating and defining as part of the Telecommunications Act of 1996 were industry changing. Now if MFS interconnected using these new interconnection agreements it would be possible for MFS to obtain access to the unbundled Copper Pairs from the Central Office of the local monopoly to go the last mile to the home or small business. This would give MFS the ability to sell dial tone or new Data services from the CO to any location in a phone company serving area using the monopoly phone company assets. A CLEC like MFS could sell ubiquitous dial tone or DSL data services because of this deregulation act. Competition in the local loop could finally happen everywhere not just where MFS had fiber. To this MFS could add some sort of dedicated modem and aggregate the data into our network at very low cost. Thus was born a quasi-skunkworks project within the company to create a dedicated last-mile solution.

A FIRST IN DSL

On May 2, 1994 MFS purchased San Francisco-based Centex Telemanagement, Inc.[205] Centex offered telecommunications to medium and small businesses, targeting a market that MFS had planned to enter using some of the $1 billion cash infusion. Centex mainly operated by reselling ILEC Centrex services. Small businesses could buy professional PBX services from someone other than the ILEC, and usually on more favorable terms.

By early 1996, MFS had absorbed the Centex business creating MFS Intelinet and brought the most talented employees on board. As the Centex business became fully absorbed by MFS, some of these former Centex people began looking for a fresh challenge within MFS.

The Telecommunications Act of 1996 became effective on February 8, 1996, and made dry copper loops accessible to service providers, by mandating wholesale access to the incumbent networks. Since the expertise of Centex was focused on reselling incumbent services to customers on a CO basis, a "skunkworks" project was developed using the dry copper loops to create a new service offering in this vein. Michael Malaga wrote a comprehensive business plan, and within MFS Datanet, I developed a technology research project with Bill Euske's encouragement, to explore alternative ways Internet users might reach their ISP. This became the first commercial DSL service offering.

The business model resembled the Centex model, in that it involved putting sales people on the street knocking on doors within about a mile of the CO and offering advanced services delivered over relatively short wires in a downtown business district. As soon as a minimum number of subscribers are sold to prove the viability of the wire center, service moved on to build-out the next one. This way the business model grows on a success basis with minimal capital outlay.

In order not to cannibalize the lucrative dial-up revenue stream, the service was focused strictly on small-business customers in downtown city areas close to the central office. This audience was already largely MFS voice customers, and so not at risk of impacting the ILEC dial-up users. It was aimed at businesses who needed a dedicated, always-on connection but lacked the budget for a T1, or fractional T1 connection. Speed was limited to 144-kbps using then available technology, but this was much preferable to dial-up, and blazingly fast for the money. A dedicated T1 could cost $1000 per month, while a dedicated 128-kbps Fractional T1 might be over $600. Our 144-kbps connection was a little over $100 per month, and cost peanuts to deliver, thus making a nice profit. The economics were compelling. Over the coming months, this service was rolled out to 53 CO and was by any measure a success. Many small businesses started with our little "skunkworks" lines, then upgraded to a T1 or even a LAN connection as they grew.

---

[205]http://www.bloomberg.com/research/stocks/private/snapshot.asp?privcapId=100554

WORLDCOM BUYS MFS

August 26, 1996, WorldCom bought MFS for $14.4 billion dollars.[206] A clear contributor to this valuation was the massive stream of ILEC revenue from the dial-up ISP customers. The sale represented a tremendous return for PKS and the public shareholders. PKS had invested approximately $500 million and held a bit over 50 percent of the shares. They received around $7 billion for their investment.

Private Networks Decline

Data communications had heretofore always been the province of private Enterprise and government, closed networks engineered to support a singular entity. Tymnet and Telenet had served a similar community with a shared networking environment but nonetheless functioned very much as a private network for Enterprise and government customers. Even the Internet – as fostered under the auspices of the NSF – was very much a private network, serving a tightly restricted community of academics and researchers.

The rise of the Commercial Internet created a tectonic shift in networking. Suddenly, networking was a commodity, available at low cost to the unwashed masses. The concept of private voice networks, central to the original revenue model of MFS and WorldCom, became almost non-existent by January 1997. The concept of private data networking became lost in the Internet gold rush.

A private-network focused business unit like MFS Datanet had been ceased to be of interest, not just to WorldCom, but to any of the big carriers. Recognition of the differences between a private network and the Commercial Internet vanished.

NorthPoint DSL Spins off

Soon thereafter, WorldCom became aware of the DSL service. We can only surmise they were horrified at the prospect of undermining that beautiful stream of dial-up revenue flowing from the ILEC. We had gone to great lengths to position it so it did not do so, but that didn't seem to matter.

Whether that was partly the reason or not, management quickly declared that they did not wish to continue the product. They fired the entire team responsible and withdrew the product. Those involved, including me, found ourselves unemployed, with money in our pockets from our vested MFS stock options, and a business plan for a DSL CLEC, along with the experience of having built the first Commercial DSL Network in the world.

Thus, *NorthPoint Communications Company* was born. I, along with Michael Malaga and the rest of the team became founders of a fresh new start-up of our own. NorthPoint Communications launched in June 1997, offering DSL services to subscribers, and high-speed aggregation of those always-on DSL connections into ISP backbone routers. NorthPoint would not be so lucky competing against the ILECs, but that's another story for another book.

---

[206] http://www.wsj.com/articles/SB841024262424406500

POST-DATANET INNOVATIONS

After the dismantling of MFS Datanet, Scott spent the rest of 1996 as VP of Strategic Development. He was a department of one assigned to report to Kathy Perone, out of New York. She had built the New York City Network and salesforce for MFS Telecom. She was a great leader and very savvy.

Scott had a good relationship with her and she had watched him build the MFS Datanet sales organization that did compete with her Telecom sales force for customer mindshare. They had mutual respect and he felt she was open-minded and supportive of his ideas. He decided to work on two projects that were potentially related or could be discrete depending on how things evolved.

Scott was supposed to think strategically which was also a way to get him out of the everyday production of revenue and let him think long term. He had been watching all the kinds of connections we sold to customers from the different MFS companies.

He decided that there was a need to reduce the cost of pulling cables from the MFS PoP in the building to the customer premise of each Enterprise in the building. There was a need for an Enterprise connection to the private Enterprise backbone, then a need for multiple Information Service Provider connections and by now there was a need for a connection to the Internet.

Of course, there were still T1 local loop circuits sold by MFS Telecom to connect to the long-distance network and multiple local loops to bring dial tone services to the Enterprise PBX. In some cases, there were analog video links to connect as well. He saw each one of those connections as a revenue opportunity for MFS but was aware of the cost to pull those cables inside the building riser.

Scott decided to leverage the experience of the streaming media services for NBC Desktop News, the applications we enabled for large customers like JP Morgan, and the need for an Internet connection to every Enterprise in every building.

Scott had used his Sales Automation Tool (Saratoga) who Patrick Sullivan had integrated onto the private MFS Datanet backbone, to look at quotations for private networks all over the USA for private Enterprise and Content Provider network opportunities. He had learned through lost sales to these companies by evaluating the input from the customers via the sales force, that the local loop cost often added up to over half the total monthly recurring cost of a private network. It became obvious that there was a need for a shared infrastructure of local loops via routers and distributed servers just like they had done for NBC Desktop News but shared among several companies not dedicated to one company to spread the costs among all the applications delivered to users.

Riser Project

 Scott was convinced the cost of the cabling inside the building was a huge sales barrier and slowed down installed revenues since he had seen it first hand at MFS over the years. Kathy Perone had seen it also and understood the idea when he presented it to her. He hired a young sales representative who had worked for Houston Wire and Cable, the  distributor of Berk-Tek Plenum cable over 10 years before for what was to be a 3-month project. Kathy Perone allowed Scott to budget for him as a consultant for a study on the internal costs of pulling cable inside the riser. Scott had him obtain pricing from a contractor that he knew from his YSA days in Houston to establish the costs associated with installing and owning those cables in MFS buildings.

 The study showed that MFS was spending millions each year on these sunk costs and there was no end in sight. The costs in NY City were significantly higher than in Houston and Kathy helped document those costs with contractors out of NY.

The young consultant's name was Grant Robertson and he did great work creating the documentation for this project. Scott wrote a summary email and  submitted the report to Kathy. It showed that MFS was spending lots of money needlessly. MFS could save a lot of money by installing a fiber optic backbone in the riser and delivering services over a shared fiber backbone. Kathy understood the concept and did sponsor the report up the chain inside of MFS. (See the strategy email[207] to Kathy and the Riser Report).[208][209][210] It never caught on inside of MFS.

RiserCorp Spins off

The idea was abandoned by MFS. Scott's friend Mike Schmitt, who was an independent contractor, asked if he could take a copy of the report and promote the idea as a business. Scott agreed and wished him well. Mike took the concept, the report and the cost infrastructure. In 1997 and started a company initially called RiserCorp, but it became Allied Riser Communications Corporation by the time of their IPO in October 1999.[211] Their stock shot up but then the Tech Stock crash in April 2000 stunted their success. The remains of the company were then acquired by Cogent Communications in February 2002.[212]

---

[207] http://tinyurl.com/RoutingPaths-RiserStrategy

[208] http://tinyurl.com/RoutingPaths-RiserReport-1

[209] http://tinyurl.com/RoutingPaths-RiserReport-2

[210] http://tinyurl.com/RoutingPaths-RiserReport-3

[211] http://www.nasdaq.com/markets/ipos/company/allied-riser-communications-corp-7095-2850

[212] http://edgar.secdatabase.com/2955/91205702003942/filing-main.htm

Metered Application Service (MAS)

During this post-Datanet time within MFS Scott promoted the concept of a distributed server service that would be deployed across the entire MFS infrastructure on a national basis.

Scott had been working with Rex Shelby, David Berberian and Larry Ciscon of Modulus Technologies for several years since the "hoot-n-holler" voice application over the MFS national Ethernet for a customer RW Smith. Cynthia Bissett Germanotta had attempted to get the "hoot-n-holler" services in front of NBC and Enterprise customers in New York, generating some interest. Inside of MFS, though, various people resisted. These people were less impressed with the viability of promoting voice over Ethernet using TCP/IP despite the fact Scott and Cynthia had deployed working versions from Scott's desktop in Houston to both Chicago and New York.

Scott describes developing the Metered Application Service model:

[---]

We had voice over IP working as an application service before anyone had heard of the idea. The phone-heads inside of MFS who sold long-distance and local dial tone hated the idea and thought it was totally crazy. It did not take off but the idea of application services was something I continued to pursue even though this application was being laughed at internally.

I did not give up on the bigger idea of creating application layer services but decided to focus on applications that did not run well on the Internet and were too expensive to deploy via a private network for one Enterprise.

I was thinking about a distributed server architecture to put the application close to the end user by having a dedicated backbone for the application services and distributed servers in all the buildings across the USA that MFS owned. I went to Microsoft and tried to get them to understand that we needed to have a way to have QoS to the desktop by application. This was a crazy idea in 1996. When MFS was purchased by WorldCom in Jan 1997, I decided to keep working on the idea and promote it internally in WorldCom. I still worked for Kathy Perone in 1997 and she allowed me to stay strategic in thinking so I kept pursuing this idea. Microsoft wanted a name for the concept and a code name for the project. I called it Metered Application Services or MAS for short. It was clear what it was about. I wanted to meter applications like streaming media or video conferencing to the desktop and sell the application as a service versus selling bandwidth or circuits. That was a very crazy idea in 1997/98 but I worked it up inside of MFS/WorldCom.

John Griebling introduced me to a small group that had been purchased by WorldCom called GridNet out of Atlanta so I decided to work with them and team up resources. I hired an ex-sales representative from MFS Datanet as a consultant and he helped me write the business plan. I had to use my financial model I had paid for in 1996 that allowed me to model a monthly recurring business. I purchased a backbone internally from WorldCom and local loops in the business model to see if it made money. I wanted to know if the business could make money buying services internally using standard carrier pricing.

The model I developed showed that it did make money as an application service that was metered. It made a lot of money so I wanted to promote the idea inside of WorldCom and get funding for it like a startup inside of WorldCom. It was the extension of MFS Datanet or the next version where we moved up the OSI model from Layer 2 or 3 all the way to Layer 7 which was the application layer.

The MAS project was written up out of my little budget and I paid all the costs of developing the business plan from this strategic budget inside of WorldCom. Kathy Perone supported my efforts and allowed me to drive the project entirely on my own. Of course, John Griebling was a huge help as were the people at Gridnet.

I needed some help so I hired Bill Collins as a consultant. He was without a job since the business he had worked on for the last few years had failed. He had convinced some people to copy the MFS Datanet Streaming Media Services product we had developed in 1994 and try to make it a business. It failed and Bill told me he was wandering around Seattle like a homeless person without a job when I gave him a chance to be part of my ideas again. I felt streaming media as a service was just one service you might have, but it was not a big enough market to be a complete stand-alone business. I think the failure of his prior business proved I was right, it was not a stand-alone business.

Bill Collins was one of my ex-MFS sales representatives and I had put him on the NBC account when I could hire a direct sales force. Bill contributed to the streaming media services by listening to the customer and collaborating with me and John Griebling to get the services developed we needed to meet those customer needs. John was the person who had the technical expertise to drive the organization to develop the services needed to stream content to the PC desktop in 1994 but Bill had been educated and exposed to the content opportunities on the MFS Datanet dime.

Since I knew Bill Collins understood my ideas and agreed with them, I decided to hire him as a consultant. I told him if we got it funded I would try to get him hired as a part of the team. I paid him a lot of money out of my budget and he was very appreciative to be able to work on my idea of application services with streaming media being one of those applications. He did a good job of researching the market and helped write the narrative. I had to do all the work on the financial model since Bill did not seem to understand what all the costs would be and how much we had to charge to make money. I had to make sure the business could make money deploying application servers on top of an IP backbone dedicated to just selling application services via a dedicated local loop and national backbone.

I was excited that I could demonstrate this in a financial model that a financial person could understand. I had paid $25,000 of my own money to a CPA firm to design and build me that model and I had laid out how it needed to work based on what Sunit Patel had taught me when we had negotiated for the value of MFS Houston years before.

One idea I understood extremely well and had to make sure both John and Bill understood in our planning process; It was clear that the business could not afford to pay the local loop "last mile" cost if there was only one service like streaming media to the end user customer location. You had to share the local loop cost across multiple application services or you could not afford to have broadband connectivity for every end user.

This was something I discovered through my experience at MFS Datanet where we bid single application, special purpose backbones to big content providers or big Enterprise customers. I had over 60 dedicated sales reps and 40 dedicated Network Sales Consultants at MFS Datanet so I was exposed to a lot of projects that would get killed because the local loop connection costs were too high for just one application. The use of the Saratoga Sales Force CRM tool that Patrick Sullivan our CTO had deployed had given me visibility no one had. It was why I was so convinced this would work financially when the network and computing infrastructure was shared.

This was a basic premise I understood that had to be part of the logic for a shared server, shared but dedicated backbone service and shared local loop to the building to make the application work financially versus just work technically.

Special Purpose Virtual Private Network or (SPVPN)

I had to come up with a name to promote this idea internally. John, Bill and I talked about it and settled on calling it a special purpose virtual private network (SPVPN). The concept of a virtual private network (VPN) was becoming something people wanted and carriers were starting to offer. Offering VPN services in lieu of dedicated circuit based backbones like T1 or DS3 networks was becoming common and well-understood.

It was clear that the applications like streaming services on dedicated servers or applications like desktop video conferencing services had to be on this dedicated high Quality of Service or high QoS backbone. I understood what would happen when a company managed the backbone the way Nathan Gregory had done for MFS Datanet versus the Internet. It would facilitate these applications to be delivered to each Enterprise customer in a VPN way. So, this application service delivered via a VPN needed a descriptive name that sounded like a carrier product.

Special Purpose Virtual Private Networks or SPVPN became the name of the business plan. This was in lieu of the name I had used before which was *Metered Application Services* or MAS. The guys at GridNet had been purchased by WorldCom the year before so they felt the VPN buzzword was better than the crazy idea of applications which carriers had no idea what applications were.

I put Bill Collins name on the front cover of the business plan since he had helped a lot with the language of the written document and I knew it would help him get a job if this did not work out. I also decided I would have him help me pitch the idea internally at WorldCom. (See the SPVPN business plan and financials to see what the concepts were and how we presented it internally at WorldCom.

Bill Collins was completely appreciative of the respect I was showing him by bringing in him on my project and listening to his input on how to communicate my ideas, paying him a lot of money, putting his name on the document and pulling him out of a mental and financial ditch.

CompuServe was an information service provider that had a proprietary backbone that was not the Internet. WorldCom bought them and closed on the deal in 1998. I was told by people in Tulsa who were ex-WilTel people that they liked the idea but Data Services were not important. One who was interested but very busy with another project for voice services said data services revenues were only $1 billion dollars per year. He said long-distance was $30 billion per year so they did not care about a startup plan for application services that might only be a few hundred million per year. These application layer services were not going to be big enough. I pointed out there was potential to grow to be billions of dollars, there was almost no churn of customers once they signed up, and these data-oriented application services were high margin versus the long-distance business margins.

I pointed out that industry data showed data traffic was crossing over and exceeding voice traffic on networks in 1998. They said since Data was only one billion in revenue, they did not care about it. WorldCom was a long-distance company not a data services company. They said maybe I should go to Peter who just came on board to WorldCom in the CompuServe purchase. Since that was a data service maybe he would like the idea.

In March of 1998, I went to Columbus Ohio, where CompuServe was headquartered, and presented the idea of SPVPN leveraging the GridNet organization to Peter Van Camp. He was very interested and seemed impressed with the idea. He said point blank it was too early in the new CompuServe/WorldCom acquisition to do this idea. He passed on it. I was disappointed but not totally surprised. He was new to WorldCom and did not have the internal leverage to do something so radical.

The people at GridNet went to their legal department and asked if they could publish the SPVPN business case on the Internet. They said yes and put it on their website. When they did this, it became a public domain idea so it was possible to take the idea out to others. I had spent years developing this under MFS and WorldCom so I thought it had merit. I was tired of fighting battles by this time and thought when my options vested in the next month I would leave WorldCom and just focus on FYI-Net.com. Maybe I would revitalize it under FYI-Net.com but knew it was high risk and would take a lot of money.

[---]

FYI-Net.com

The year 1998 brought monumental change to Scott's life. MFS's success made his stock valuable. He wanted to start a new company. Having failed to sell his SPVPN/MAS vision to WorldCom, he chooses to join with his cousin Paul Yeager and a software developer named Massoud Rad and pursue a vision he had of how to communicate more efficiently to a target audience like a sales force over a network.

Scott had worked with his cousin Paul Yeager who was a producer for PBS and had award winning movies for PBS. The idea was to put content in a data base and have it served up dynamically out of a database for the end user based on the profile of the end user. Scott had taught sales reps all over the USA for years on what LAN's, MAN's and WAN's were. He also had taught people what the Internet was so he thought if he had a 3 D world and it was animated he could teach a large sales-force these complex networking ideas efficiently if delivered over a network. Paul Yeager who was a visual story teller agreed and he said he could do it. They found Massoud Rad who was a computer science person who had delivered streaming media over a network for a dating service within a building on a LAN.

Scott left WorldCom in April 1998 to focus on the new company. Together they were the three founding partners. They named their company *FYI-Net.com* and envisioned an innovative model of services revenue, based on a broadband content development function under Paul and a broadband software application services function under Massoud Rad.

FYI-Net produced interesting and novel work including an online Content Management/Learning Management System. It also developed and deployed a fully HIPPAA compliant medical application. Their collaboration innovated in several directions, not the least of which was a series of stellar animation videos for education and marketing. The idea was to focus on deploying a broadband-oriented content publishing and learning management services application focused on distributing broadband content to targeted audiences.

Perhaps the most monumental result of his work with FYI-Net.com came when Enron began recruiting him because they heard about him from Stan Hanks as being innovative. He still had his idea of a *Metered Application Layer Services* (MAS) developed under MFS/WorldCom and then rejected by the new management. In time this led to his joining Enron. Scott demonstrated the 3 D world and animations and taught Enron people how FYI-Net was an Application Service Provider that could use a distributed application service network infrastructure.

In his autobiographical notes, Scott explains:

[---]

I wanted to use a directory service to publish content out of a database on the fly to target audiences for FYI-Net. This would be a new kind of application service that was paid for in a subscription or pay per view way depending on the needs of the customer.

The guys at GridNet wanted to demonstrate how the WorldCom VPN services worked so FYI-Net.com received an order to develop an animation explaining these services. It was a big deal to have a 3D world to animate complex ideas and concepts so people could visualize these ideas. (as seen in this video spoofing the dot.com culture Net.com.com)[213]

Later, I decided we should develop a new broadband environment that was a CD-ROM front end to a 3D broadband world with a hot link to online content that was served up out of a database. The content served out of the database was linked to where you were in the 3D environment. This became a product we called the *Network Resource Kit* and was a way to demonstrate how a broadband environment might differ from traditional text-based websites. There was a lot of coordination between Paul and Massoud to make this happen.

The merging of content creation capabilities with online database capabilities required tremendous coordination and creativity across multiple disciplines. Of course, now we have this linkage to what you are interested in every time you do a search in Google but it is still not in a 3D world where you pick the linkage to your interest by clicking on a 3D representation of that physical object in the virtual world.

It was a pleasure and joy to see how Paul and Massoud and their teams came together to develop and deploy the applications and content for the environment. The website that was developed and deployed showed the concepts of broadband streaming content delivered via the directory driven databases in real time. The goal was to create a new kind of user experience that utilized streaming media to demonstrate a web experience that was broadband versus the text-based experience of 99% of all websites. The common and well-understood approach was to assume a dial-up connection and then minimize the amount of time spent waiting for a download of text with small graphics. At the end of 1999, the assumed connection speed was 28.8 kbps, not 500 kbps or faster.

[---]

---

[213] https://www.youtube.com/watch?v=o1rUS-o12Ms

Enron comes knocking.

In mid-1998 Scott was approached by Stan Hanks, then at Enron Communications Inc. (ECI) out of Portland, Oregon. ECI was a startup under Portland General Electric (PGE), which had been purchased by Enron Corporation. PGE, the electric utility company serving Portland west of the Willamette River, was founded in 1888 as the Willamette Falls Electric Company. Enron bought the utility company on July 1, 1997. In 1998 ECI was launched to expand the company from energy into communications.

This startup was laying fiber optic cable between cities and the founders wanted a new and innovative "big idea" to develop. Scott told them he had his own business but would talk to them as a potential consultant. He was not interested in going to work for a large company. (Enron was in the top 100 at that time.)

Scott Continues:

[---]

I told Stan my idea of *Metered Application-layer Services* (MAS) and explained how it had been turned down by WorldCom. I sent Stan an email in May of 1998 and said "Think about creating the infrastructure for end to end services. Make deals with all local providers to give them a backbone, collocation and access to off net local loops via a tariff engine and provisioning engine. Include the software that is needed to deliver high QoS end to end that is intelligent about the application and maps it to the appropriate backbone and uses any local loop approach. Think about doing a deal with Netscape and Sun to kick Microsoft in the butt and demonstrate this capability with the whole thin client side of the equation but we could flip flop and do a deal with Microsoft as well. The intelligent software would be able to work with Microsoft as well the whole java approach.

We would be agnostic but could do a deal with the java side of the equation to stir up the market and get a lot of ISP's and CLECs to want to do deals with us immediately." I included the document "*WorldCom Metered Application Services Prototype Project Definition*[214]" as an attachment.

I explained it was not covered by a non-compete or other restriction since I had developed it as part of MFS Datanet and carried it forward as part of WorldCom. I had been talking to Rex and his team from Modulus Technologies about taking the idea to others, but once I got involved with FYI-Net I had decided to focus on that instead. Now that Stan was approaching me and asking for an innovative idea I decided to tell him about the Metered Application Services idea. I had sent him an email with the MAS write-up from when we were dealing with Microsoft which included the use of the Modulus Technologies InterAgent software as the middleware layer.

---

[214] http://tinyurl.com/RoutingPaths-MetaApp97

Stan told Joe Hirko and Ken Harrison about me and my ideas. He set up a meeting between Ken, Joe, and me. They came to the FYI-Net offices and I showed them the idea of broadband content served out of a database as an application service. To illustrate this concept in early 1998 I used "The Network Resource Kit" as the broadband user interface on the computer and showed how it was linked to the online environment. The online part was not static web pages in text as most websites were at that time. The broadband user interface was linked to dynamic content served up in the context of what the user was interested in at that moment in the 3D environment. In this instance, it was linked to data about desktop computers, or Local Area Network components like Ethernet LAN switches or to router information like Bay Networks or Cisco.

I explained that this environment needed someone to offer streaming media as a service to enable a 3D broadband front end user-interface out of the network. I told them about the need for a distributed server application service and that no one who was a carrier class player was doing anything at the application layer other than email. Email was not particularly time sensitive to the user and it was all text-based, thus very small. I told them even email needed to have PowerPoint or user made movies attached and this was going to make it behave more like a broadband application over time.

I said I would help them develop these ideas for a business as a consultant but I had already developed the concepts, costs and revenue models for all this as a project at MFS and then later for WorldCom. They said they wanted me to come to work full time. I told them "no" and said I did not want to work for a big company like a power utility or Enron. They came back to me two more times and said I should consider what it would take to do this.

Level 3 was getting going as the next Jim Crowe project, Royce Holland had started Allegiance, Time-Warner Telecom had hired Larissa Herda and I had consulted with her about Ethernet services. Other ex-MFS people like Kathy Perone had started DSL or hosting companies. Pete Farris who was over John Hardie and the sales people in D.C. that handled MAE-East was going with a startup called Equinix and Mike Schmitt my friend since 5th grade who worked for me at YSA had taken my riser cabling infrastructure idea and created Allied Riser Corp. The year 1998 was full of startups who were raising money and going public or intended to go public in 1999.

I thought that if a big company like Enron would put the money in the *Metered Application Services* (SPVPN) idea/deal and not go public initially, then you could build a business, get revenues and have a chance to make it a profitable business without the crazy expectations of going public. Joe and Ken wanted to go public and spin out of Enron but we would have the Enron umbrella to get started. I hoped we would not go public and have Enron money to build the business so we did not have the same scrutiny of being a public company. It had been a difficult time at MFS when we went public to always have to make your numbers every month and they seemed unrealistic so that concerned me. If Joe and Ken could get Enron to allow *Portland General Electric* (PGE) to fund it but leave us alone and be under Joe and Ken's control, then it seemed fun and challenging which was enticing.

The next week after a long soul-searching weekend with my family, I met with Joe and Ken and told them ECI had to develop this new application services business. They could not be in the long-distance, dial tone or ISP business. They could not go out and be a web hosting company, be a DSL provider or be a local loop CLEC like MFS.

If they would commit to this new type of business that was only focused on selling application services to small, medium or large Enterprise customers and go after the content sources to convince them to put their analog content out on the Enron Network to deliver a new type of broadband content to the computers and TV's using the TCP/IP protocol to route the content to the business end users or consumer end users. We would agree to leverage the existing metro networks and deliver it over Ethernet connections to the business and consumer end users. If Enron Communications would agree to these radical ideas in 1998, then we would make the sacrifice to our family time.

I told them I wanted to be in the role of VP of Strategy and cross over all the boundaries in the company to make sure all the different cultures of software applications development and server deployment meshed with the fiber optic transmission, and TCP/IP routing cultures. Ken and Joe wanted me to be CEO and I said that was not my strength.

I was not going to be CEO or VP of Sales or over any operations or engineering. My job was to help build a collaborative culture to make this new kind of company work. I told them the right kind of culture was as important as any specific technical capability and they had to commit to building the right culture so we could succeed. I explained how MFS had a circuit based point-to-point culture for metro services which was great but Ethernet Services and LANs, in general, were by definition for getting information and data to anyone on the LAN. This "any to any" nature of Ethernet required a different mindset and culture from circuits but closed user group or Enterprise private networks was different from the Internet world which was anyone on the planet earth getting to anyone else using global connectivity.

This was also a different way of thinking. Finally, software developers then were mostly thinking Client/Server technology over a Local Area Network or LAN and they tended to not pay attention to or care about the underlying network especially the Metro Area Network (MAN) connected to a Wide Area Network (WAN). Software developers were culturally different than Circuit Fiber people, who were different from Layer 2 Ethernet LAN people who were different from ISP or TCP/IP people who were different from Application layer Server people. We had to blend these cultures together to succeed.

I suggested that to get a foothold on these cultures they had to buy Modulus Technologies to get the software development expertise and culture in the company. Also, Modulus Technologies had the InterAgent Software that could be used to let the applications talk to routers and layers below to provide tiered QoS driven by applications. This middleware was essential I felt to the plan.

I told them to hire Bill Collins as a specialist in the content services business since I had funded his experience as a sales representative in NY for MFS Datanet. He had gone out and failed at trying to copy our Streaming services from MFS we built for NBC in 1994 but he convinced me he had learned a lot, made content provider contacts and could bring customers if we had the infrastructure. I believed him.

I said we should hire Jeff Foster as a directory services expert out of Quest since he had been with a big RBOC, focused on using the Microsoft Directory to serve up content and gotten Cisco involved. FYI-Net did some animations explaining this concept for Cisco and US West/Quest to illustrate these ideas. I knew that LANs had to have QoS so they did not get congested and thought the Cisco and US West-Microsoft concept of a Directory driven QoS would catch on. Since I had shown the idea to Microsoft in 94, through 96 they had adopted the idea of application layer QoS and put money into it. Joe and Ken agreed to these requests and I decided to go for it.

[---]

Paul Yeager and Massoud Rad continued to run FYI-Net.com, focusing on large customers. Large customers possessed the marketing dollars to sell their high-tech products. FYI-Net created numerous applications but the flagship was an online Content Management/Learning Management System.[215]

FYI-Net also developed and sold a fully HIPPAA compliant medical application.[216]

Many customers were interested in using this 3D world for marketing their products, but having ECI be an early adopter funded the 3D server farms and workstations needed to create advanced animations illustrating complex business concepts. Animation was much cheaper than filming an actual scene using actors and the 3D world supported concepts that did not exist in the physical world. Marketing people at that time. Marketers of the day, focused on using text-based websites, had trouble grasping the potential.

---

[215] https://youtu.be/HdS0NEAUReE
[216] https://youtu.be/rTc_xLKtHqM

Enron Communications Inc. (ECI)

In mid-1998, Joe Hirko and Ken Harrison decided that Joe would become the CEO of *Enron Communications Inc.* (ECI) and it would be run out of Portland where PGE was located. They picked Steve Elliott to be the CFO since he had worked for Joe at PGE. Joe had been the CFO of PGE and he had been involved in the negotiations for the purchase of PGE by Enron. Ken Harrison was the Chairman of PGE when purchased by Enron. Joe and Ken had to agree to hire Scott and agree to meet his requirements or there would be no deal.

Scott later discovered that Ken and Joe went to Ken Lay to check him out more. Ken Lay — who Scott did not know — called Scott's friend and mentor Jack Trotter dating from his time with *Houston Lighting & Power* (HL&P) in 1987. Mr. Trotter later said that he told Ken Lay there was no way they could get Scott. Trotter said Scott was a pure entrepreneur who was not interested in working for a big company, especially one as big as Enron.

Mr. Trotter confirmed Scott's past accomplishments, saying they only motivated Mr. Lay to want him more. In Scott's notes from the period he describes events:

[---]

All this happened when I was turning down Joe and Ken Harrison and saying I would just be a consultant. Ken Lay told Mr. Trotter that he would get me and to "watch him." I had no idea Mr. Trotter had spoken on my behalf and had told the truth that I was not inclined to work for a big company again.

Once Joe and Ken Harrison talked to Ken Lay they had the nod to get me by agreeing to my requests. Once I knew they would build a business focused on using a private fiber optic backbone to develop and deploy application services and not sell raw bandwidth or connections to the standard nondeterministic Internet then I felt I could commit to this new company. Meeting my request for equity as options was a big plus but only if we succeeded so my focus was on making ECI successful and changing the entire world around what was possible and would be expected from a broadband environment.

I became a consultant immediately and went up to Portland with my financial models. Joe hired Bill Collins and Jeff Foster first and they started working on the business plan based on the concepts in the SPVPN plan from WorldCom which was based on the MAS efforts at MFS Datanet and WorldCom funded and evangelized out of my budget as VP of Strategic Development at MFS and later WorldCom. Modulus Technologies had been one of the players I felt was important to the success of managing and deploying metered application services.

I wanted ECI to purchase Modulus Technologies as well so I was not an employee yet. During this time, I injured my back with a herniated disc. It was extremely painful and put me on my back literally for over a month. I tried to conduct meetings over the phone while on pain killers but was not very effective. I directed Bill and Jeff to get things going on costing out the business with John who worked for Steve.

I thought that since I had gotten them both equity at ECI that was well above the level they would have received otherwise, that they would be loyal to me while I was on my back.

I became an employee in the September time frame after some procedures to my back that helped me not be in total pain all the time. I held a planning meeting with Jeff Foster and Bill Collins in a conference room and we came up with 6 products that we would develop and deploy. We started to define those products and how we would create them as services.

[---]

Joe and his team were busy during 1998 building fiber from Portland to Los Angeles and swapping or building a return route. They had a full-blown fiber optic construction effort that was very sophisticated and successful at building fiber routes with either 144 fibers or 218 fibers in the cable jacket. Scott was not involved in that construction and did not bring anything to the table around that endeavor. Joe and his Portland Team did an amazing job of building these routes which was also what Level 3 and others were doing. ECI was doing this with one group while developing the applications services with another.

Scott was focused on creating a logical national backbone without being dependent on the specific fiber installed to create a dedicated IP backbone not connected to the Internet. This private IP backbone would allow ECI to deploy application servers and sell those applications on a usage basis yet have high Quality  of Service because ECI controlled the traffic on the private network. Scott discussed the ideas with Joe and he put out a memo[217] to all employees that defined how ECI would be different culturally as well as technically.

"This is the initial step in creating a new type of Business/Product Development process that will allow EC to respond to the market demands rapidly with the correct products. Ultimately, this will become a strategic advantage for EC more than our IP backbone, fiber in the ground or any individual application service.

ECI intends to develop some strategic capabilities in software integration and universal messaging to offer these application services. Modulus Technologies (Rex Shelby and Dave Berbarian) has developed some software that can allow EC to communicate with software developed by Microsoft, SUN, Oracle, People Soft, Netscape and even that little company IBM, as well as many others. This capability is critical to developing and offering these services. EC will own the unlimited use of the necessary software license required for this suite of capabilities since this is a strategic asset.

Other companies will become software vendors or subcontractors that will assist in the development and integration of the suite of software needed to roll out these services. This software integration and universal messaging bus capability will become a core competence of EC and a strategic capability necessary to make EC a sustainable business."

---

[217] http://tinyurl.com/RoutingPaths-EnronVision

Scott has seen the culture of a circuit oriented, point-to-point thinking, local loop company; he knew from MFS the difficulty it created getting a LAN based service offering going. He had seen the WorldCom people reject the idea of a Metered Application Service because they were a long-distance business and their culture focused on selling long-distance minutes. Joe Hirko agreed that enlisting the company culture was critical to creating a new category of service offering within an application layered service company, which spanned the OSI Model layers from Circuits to TCP/IP.

National Association of Broadcasters

In January of 1999, they opted to focus on two of the six products to make sure those were developed, deployed and ready for the *National Association of Broadcasters* (NAB) conference in April of 1999. They intended to demonstrate these first applications on their dedicated fiber infrastructure and demonstrate them in front of the content creators attending the NAB trade show. There was a belief in the industry that TCP/IP was not able to deliver analog video quality at high bandwidths like 270 Mbps all the way down to 8 Mbps. To deliver high quality video to the desktop computer at LAN speeds for Real Networks or Microsoft Streaming clients, including variable data rates from 28 kbps up to 2 Mbps, was a new concept for a service provider. These were radical ideas in late 1998 and early 1999. The demonstrations of these capabilities in April 1999 at NAB was an industry shaking idea.

The two Products addressed were:
- ePowered Media Cast™
    - High-Quality Video Streaming for Live, On-Demand, & Scheduled Events
    - 10 Times Faster than Internet, Large high-resolutionViewing Window, and Stereo Sound
    - Ad-Based, Subscription or Pay-for-View Contracts
    - We had FYI-Net create animations that explained this service.
        - The first presents from the perspective of an ISP partner[218] and
        - The second presents from the perspective of a Content Provider[219] partner.
- ePowered Media Transport™
    - Flexible, Affordable Broadcast TV Transport
    - Provides Alternative to Satellite and Land-Based Networks
    - Usage-Based Contracts
    - We had a different animation developed to explain how this product worked as well.[220]

---

[218] https://youtu.be/ASwCKLXGRw0
[219] https://youtu.be/WaRmr2T4T8o
[220] http://www.youtube.com/watch?v=ASwCKLXGRw0

ECI went all out to deploy a first-class booth at the trade show, one that demonstrated both the video streaming services and the Media Transport services, sparing no expense to make a splash. A video that was professionally developed by the marketing people in Portland ran in the booth as a video loop.[221] People were standing three or more deep in front of the multiple screens in the booth for the presentations.

There were sessions at the ECI booth and at the Real Networks Booth.[222] They had tremendous traffic at both booths and made a huge splash at the NAB show. They also had a presentation at the Sun Microsystems Booth at NAB.[223] The traffic at all three created a huge splash. The content creators and aggregators at the NAB conference were very interested in seeing how this medium could be used to monetize content in the form of advertising revenues, Subscription-based revenues, and pay-per-view (or play) revenues.

Until now, all video had been analog video. Now ECI was showcasing high QoS IP protocol video services for the processing and distribution of streaming video all the way to the desktop. ECI displayed a video loop[224] showing how and why they were changing the way content would be delivered.

The demonstrations proposed business-to-business customer potential as well as business-to-consumer potential for all the content creators, who could now see how their high production value content could be moved around as digital files and streamed out to the edges of the Internet via this new Enron Intelligent Network.

Stan Hanks put out an internal memo with a map[225] showing the footprint of the EIN by April of 1999. The backbone was defined by Stan Hanks just before NAB to the employees.

Matt Harris put out an email congratulating everyone on the actual working apps over the EIN backbone.

MattH41699Media Cast Deployed!.pdf

Others circulated test results showing the traffic running on the backbone before NAB, including GIF files showing the traffic on the backbone in April of 1999.

990414_JRowhShelbyPalmer» Network testing.pdf
http://tinyurl.com/RoutingPaths-EIN-SJC-traffic
http://tinyurl.com/RoutingPaths-EIN-PDX-traffic
http://tinyurl.com/RoutingPaths-EIN-ORD-traffic

---

[221] http://www.youtube.com/watch?v=Fb1aPG5aDm4)
[222] http://tinyurl.com/RoutingPaths-RealNetworksPPT
[223] http://tinyurl.com/RoutingPaths-SUN-PPT
[224] https://youtu.be/LIm_1loK5Lc
[225] http://tinyurl.com/RoutingPaths-EIN-Map

It was quite impressive, especially when you realize the idea for this business model had not occurred to Joe Hirko and Ken Harrison until mid-1998 when Scott explained it to them at FYI-Net offices and showed them streaming animations in his offices over the LAN in the building.

Stan Hanks was interviewed by Gordon Cook who then published an excellent explanation of the new Enron Communications plan in the Cook Report for April 1999.[226]

In the interview, Stan discussed his QoS paper which explained[227] how ECI was redefining QoS and addressed the way ECI was approaching the application layer. It was also made clear QoS was a virtual non-issue if the network was not massively oversubscribed.

The team of people in Portland who worked so hard to pull this off, made a great internal PowerPoint[228] presentation that illustrates the excitement, the energy and the number of people committed to explaining these new ideas and concepts. It also shows the demonstration areas in the booths set up for Media Cast (the streaming services) and Media Transport (the real time and scheduled high quality compressed video for live or delayed broadcasts) as well as the booth staffing and energized audience. The show had been a true team effort and a huge success and Scott felt fortunate to be a part of such an impressive product launch.

ECI on the Radar

The huge NAB success put Enron Communications on the radar of all the content originators with a digital TCP/IP Solution for moving content. The capability instantly obsoleted the traditional Analog video way of doing business, the industry standard up to this point.

Scott had succeeded in getting Joe Hirko and his management team including Rex Shelby and David Berbarian, and Larry Cisco of Modulus Technologies to agree it was important to instill the culture needed for creating application layer services. A new standard had been created in the minds of the content originators about the power of creating and distributing TCP/IP packetized video. Now the job of leveraging business with it began.

After NAB, the work necessary to turn these new services into revenue generating flows started, creating services that could be purchased and deployed by real-world customers. Of course, there were trade-off decisions about capabilities and customer needs along with the usual product development disagreements internally. Securing consensus and establishing the tradeoffs necessary in the development processes took many forms.[229]

---

[226] http://tinyurl.com/RoutingPaths-CookReportApril99

[227] http://tinyurl.com/RoutingPaths-Hanks-QoS

[228] http://tinyurl.com/RoutingPaths-NAB-Show-1999

[229] http://tinyurl.com/RoutingPaths-DevProc

Enron 'Buys' ECI

Joe Hirko and Ken Harrison had originally thought in terms of ECI going public. In June 1999 Hirko, Rex Shelby, and Steve Elliott presented the Enron Board of Directors a plan to grow the company fast and to raise money in the public markets. Joe, as CEO of ECI, and Steve Elliott as CFO were known at Enron and by the Board of directors, and looked to as the ones who had to produce. One of the board members, John Duncan, asked Scott if he thought ECI could develop these products and services. Scott's faith and enthusiasm helped to sell the vision; he believed they had been on their way to success but it was going to be a lot of work.

Skilling listened intently and made a monumental decision based on what he was hearing. He decided to bring ECI inside of Enron and make it core to the future of the business. This meant Enron was committed to the broadband business long term.

This also meant that ECI did not go public. Employees were bought out of their ECI options for a combination of cash, Enron options worth nothing on that day, and stock in Enron, all vested over three years. Many ECI employees were angry because they did not want to be part of Enron.

From one perspective, the decision was better than going public because there was upside in the parent Enron, and less pressure from the bottom-line scrutiny a public corporation would face. ECI could build the company using money from Enron, who could afford to be more patient than public investors. It seemed less risky and provided the runway needed to develop the business. This transaction occurred in July 1999.

Ken Rice from Enron was made Co-CEO of ECI with Joe Hirko. There was confusion about the ECI plan since Tom Gros was working on bandwidth trading as well as streaming video and application services. This confusion was an issue so internal meetings did occur to try to get everyone on the same page.

August Internal ECI Meetings

Things moved fast. In August of 1999 internal ECI meetings in Portland explained existing products, the services and the capabilities to all the employees. Joe and Ken explained the plan was essentially unchanged for creating the applications platform called the Enron Intelligent Network (EIN). Claudia Johnson was managing PR and they had decided to brand the EIN and brand the middleware software called Inter Agent from Modulus.

The following videos and documents are some of the extensive supporting documentation provided on YouTube and the supporting website.

- Kirk Wright presents the status of Media Cast the streaming media service. Notice it shows what already works in August 1999 and what is on the development track for delivery by the end of the year. There is zero possible confusion as to what is real, and what is yet to be built. https://youtu.be/vZW1MJy1vv8

- Jim Rowh presents the status of the Network Operations and Control (NOC) and systems that are part of the intelligence of the Network to control the applications servers, the routers and the transmission systems. Again, there is no ambiguity between that existing and that which is promised. https://youtu.be/0gJvlu4FOBU

- Holly Bradford Nelson presents the status of the Extranet website where the ISP and carrier partners can go and sign up and see a demo of the real streaming services delivered over the EIN. It was a demo of the actual site and showed real streaming content from August 1999. https://youtu.be/TthYHSyLUsg

- Joe Hirko and Ken Rice are Co-CEO's by now and they explain a lot of things that are confusing about the focus of the company now that ECI has become "core" to Enron. https://youtu.be/D5rr1BKe9Qg

- Jeanette Buse shows a live demo of the Media Transport product and explains how the customer can choose the QoS, select the bandwidth and schedule it for the high-speed high quality video that is available over the EIN. The service is for either live video feeds or scheduled time delayed. This is a user defined application service and the PowerPoint is from the real application screen shots. (the audio is missing so you have to just watch her do the demo) https://youtu.be/cFm4ZffkX_4

- Merat Bagha – VP of Technical Sales – explains the order entry process and shows how far ECI has come in the productization of the services. https://youtu.be/dfiaXpANSvg

- ECOMS system is explained, showing how services are ordered and tracked internally to provide customer services for ePowered Media Cast and ePowered Media Transport. https://youtu.be/gZEuz4RuH-g

- Claudia Johnson – VP of Corporate Marketing and Communications – explains her role as head of PR and the progress made since April in defining the product and messaging. https://youtu.be/SCosDBacopQ

- F Scott Yeager explains the EIN concepts, the new EIN product strategy as of August 99, and the progress made since the April NAB Conference. https://youtu.be/6pwxKhJOWdM

- Rex Shelby explains role of InterAgent software. https://youtu.be/FOxEu_rFTPg

Bandwidth Trading

Now that ECI was core to Enron, the concept of bandwidth futures trading as initiated by Tom Gros became a more serious initiative. Futures and options have existed since the 1800s to allow the parties to hedge their positions. If you own 10,000 gallons of fuel that cost you $2 per gallon, you can sell a future contract to deliver 5,000 gallons at $4 per gallon to the buyer at some future date. The contract is sold today; the fuel is retained until the future date. You'll make more money on the remaining 5,000 gallons if the price goes up; you at least won't lose money if it goes down. Meanwhile, you received money for the contract itself.

That contract itself becomes a fungible instrument that itself has value and may be bought and sold in markets, and the value of which can change dramatically if the price of the underlying commodity changes. In our example, if the price of fuel soars to $5 a gallon, then a contract to buy at $4 a gallon is valuable.

Thanks to investments in building fiber networks and using spare fiber for swapping for more routes, ECI owned or controlled about 18,000 route miles of bandwidth to cover the USA. More importantly there was fiber from all the other carriers who were spending billions on building fiber in the long-haul and in the local loop market. The plan was to leverage all this to have a larger footprint for the least amount of money invested by anyone.

Why build when you could acquire bandwidth from others for cheap and the price seemed to keep going down with the glut that was being built by Level 3, Global Crossing and others? Scott agreed with this and since he knew MFS players who were at all the other carriers and he understood how to interconnect at low cost. He showed Joe and his team how to make the footprint huge with relatively little capital expenditure. This was a key concept in his business model since it bought bandwidth at market prices even if they used their own internal fiber. The MAS business plan called SPVPN had contemplated buying bandwidth, local loops, co-location and all telecom services at market prices. This has been demonstrated successfully to Joe and Enron or they would never have funded the business.

Bandwidth trading fit this model.

 Tom Gros, with input from Stan Hanks, had decided to trade DS3 Equivalents on standard contracts called DS3 months.[230] Since TDM circuits are deterministic it was possible to trade them. One DS3 was the same no matter who the TDM equipment manufacturer was or how many other DS3s shared the fiber. Scott had explained this to Tom Gros early on since this was something he knew well from MFS.

Futures trading became a method to hedge that bet in case the expected demand did not arise. Commodity trading is a complicated business. Tom Gros explains how it would work and shows companies how it will be possible via an online trading application.

 Tom Gros also developed a very good video explaining the way the Pooling Point would work.[231] The idea was to Interconnect the Enron "Pooling Points" bandwidth manager switches to all the other carriers in Carrier Hotel buildings and then have access to all those other fiber routes and fiber bandwidth providers. An animated video narrated by Scott explains this as well as the Tom Gros video.[232]

[230] https://youtu.be/huI5d8EtmrI
[231] https://youtu.be/k7_4Xwxc3dk
[232] https://youtu.be/OyZWuyvgvsY

2000 Analyst's Conference

Each year in January, Enron held a regular Analyst's Conference, and serious effort went into developing a presentation for the January 20th, 2000 event. McKinsey and Company, the global management consulting company where Enron President and Chief Operating Officer Jeffrey Skilling had worked, was the driver of the financials that were to be presented. The consulting company insisted that Enron revisit and justify everything.

Scott was invited to attend the Analyst's presentation. As he was writing a Strategic Vision Document to use to explain the Applications layer services, the Streaming applications as examples of the way the EIN enabled applications and how bandwidth trading and Intermediation all fit together, he felt it was important to hear first-hand how it was going to be presented. He was not involved in the development of the deal between Modulus Technologies and Sun Microsystems, Inc. though he knew they were working on licensing InterAgent as the Sun Java Messaging software or middleware layer. He wanted to hear first-hand how that had occurred and how it would be presented to the public.

While Joe and his team in Portland believed they were running ECI, it now became obvious that Houston was taking over.

Scott observes:

[---]

This is when I started noticing there was a fundamental difference in how "Enron" people looked at new businesses and how I had developed from CECO to YSA to NCI to MFS and MFS Datanet. Enron thought about everything in a transactional way first and mostly financially. The business was the transaction. It did not matter what "it" was. Gas Pipelines delivering natural gas, Electric Power Plants delivering electricity, financial transactions around trading all of these power related commodities or hedging all of this in a sophisticated series of trades. They had no emotional attachment to any business or any customer at all.

The way other companies and my experience evolved was totally different. You were in business to get revenues from providing goods and services. Being the best at making the product (CECO or YSA) to being the best at providing the Service (NCI, MFS, MFS Datanet) is the way you got the order from the customer to get the revenue every month. It started with being the best at providing the product or service, then selling it, then supporting it and keeping the customer happy.

It was new to me and I wondered if I was out of date and my way of thinking was no longer relevant in the world of online trading and huge hedging of risk. I questioned whether a customer focused product development was still relevant and whether it could co-exist in this new to me Enron culture? I had to jump in with both feet and figure out how Houston Enron did business. I was figuring out that Portland General Electric (PGE) Enron (i.e. Joe Hirko and team) were not really "Big" Enron.

I was not being instructed by Joe Hirko so much now. He and Rex Shelby and others in Portland were busy building fiber and making a deal with Sun Microsystems to license the InterAgent Software to Sun. So, Ken Rice wanted my time to educate them on the plan. Others out of Houston like Kevin Howard wanted to learn what the logic was on how the business could make money. I explained how the local loop was the most expensive part of the total network to connect to end users and we needed to have multiple revenue streams per connection to cover the total cost of the network and the usage of the local loop.

The graphic illustrated how all the different types of services or "applications" that were combined over the local loop that would make the business highly profitable. In fact, over time the incremental cost of adding new applications was very low, so the margins could become very high. This application layer services business could be a very profitable business and since it was monthly recurring revenues the discounted cash flow method of evaluation would mean the valuations of the business would be very high.

 Kevin Howard spent about a week with me and developed this one-page slide that represented the entire business proposition end to end.[233] It was brilliant and showed he understood what I was saying about all the different business models going on over the same infrastructure. I felt that the new CFO and new management of ECI now understood how much money could be made off the ECI applications services business plan and the Conceptual Business plan I had presented to him and others had helped him understand how the ECI plan made money even if we bought bandwidth in the open market and did not use our own fiber.

I was reminded of having to resell the MFS High Speed LAN idea to Al Fenn to get MFS Datanet going. Then MFS bought UUNET and forgot the difference between selling closed user group private Ethernet networks and selling connections to the Internet. Later MFS had been bought by WorldCom and as a Long-distance company they forgot that MFS was a local loop company and how they were used to sell to all long-distance companies. Of course, WCOM had no clue of anything about MFS Datanet. All that was lost. Now it was obvious that Houston Big Enron was in charge and I had to go to work again selling the ECI business plans and this work with Kevin Howard was encouraging.

[---]

Scott was asked to explain several key concepts to Jeff Skilling and to create animations to explain this to employees and customers. He realized he was now an "industry expert" and a resource they would call on to provide input to the decisions they would make. Without warning, his role in the company had changed dramatically.

Scott noted that this was the point at which he stopped being asked to recommend actions or offer guidance. His activities became merely to explain concepts and ideas and what the past looked like. He was not asked to drive the organization, ceased to have any sort of management input beyond that of "explainer."

---

[233] http://tinyurl.com/RoutingPaths-BusProp

Enron Broadband Services (EBS)

The Analyst Presentation 2000 by ECI (in which the company was renamed EBS) explained what was in place and what was being developed. Scott felt the entire Analyst PowerPoint presentation was an accurate representation of the existing network and applications platform, with information about how it could evolve over time.[234]

Jeff Skilling began the session by announcing the name change to EBS. Scott listened as they explained the existing infrastructure, the existing streaming application services and pooling points via a Tom Gros video, and saw video clips of key people inside of ECI explain how well ECI/EBS was doing at executing them. He attended three days of internal training called ePowered Applications and had seen first-hand all the progress that had been made.

Seeing how much had been accomplished and how many internal battles had resulted in a common focus on was exciting. Then the Analyst Presentation made it clear what was in place and what was being rolled out.

Two video recordings of that meeting are available. Video Part 1[235] records most if the conference presentation and Video Part 2[236] covers the BOS and the question and answer period of the presentation. A video inserted in the Edited version of this latter video about the BOS by Rex Shelby was not shown on Jan 20th 2000 to the Analysts, but it was a great training tool for EBS people internally. Also, in the latter portion, Scott McNealy of Sun Microsystems is brought in and introduced by Jeff Skilling.

Scott had reason to believe the culture of Enron would allow the changes envisioned by EBS to occur. In fact, Skilling had said in the ePowered apps meetings it was "messy" to implement change, so Scott felt comfortable knowing the COO of the parent understood the difficulties in building new businesses from scratch.

Scott also was encouraged by internal presentations about how Big Enron would and could evolve as it had in the past.

A video of Ken Lay, from the All-Employees meeting of May 18, 1999, shows what Jeff Skilling and Ken Lay were telling all Enron employees about re-inventing themselves.[237] A company that had gone from $40 billion in revenues in 1998 to $60 billion in 1999 and would grow to $100 billion in 2000 conceivably would be able to reinvent itself and get into the broadband business. They had, or seemed to have had, the resources, the time and the money to do it, and they made a compelling case for success.

In early February, Jeff Skilling told Scott that it was now part of his job to train all the new people joining EBS. He did so at an internal meeting in front of about 300 employees. Scott was officially relegated to helping train new hires from other parts of Enron and imparting the new Strategic Vision of EBS's plan into the new employees.

---

[234] AC2K Broadband Services (PPT): https://youtu.be/EY3XWR9-HCs
[235] AC2K Video Part 1: https://youtu.be/EXATyj6khi8
[236] AC2K Video Part 2: https://youtu.be/r1_ni7S3Bzs
[237] May 18, 1999, Jeff Skilling, Why_believe_Enron-energy-1Mbps.mpg

Scott believed this was a good thing and he took this new role very seriously. The expectation was it would help the different groups understand their respective roles and work together.

He still held the title VP of Strategic Development so he had to get the plan in front of the key people within EBS and make sure they knew their respective roles in relation to the Strategic Plan. He and John Hay, who was responsible for training, used FYI-Net to create and host the content to educate people about EBS using a broadband content approach to educate people in a new way.

John Hay and Scott used the EIN every day to explain the EBS services internally and to customers when they came to the building. The content, including the animations, about the EBS services were part of the story telling tools used by both to explain the vision of content and applications being delivered in a new way to audiences all over the world to any device. Scott also used it with customers when on the road. They used it to show the customers that it worked and they could test it from anywhere in the country by going to the EBS Broadband web site and viewing the streaming media that explained how the EIN worked, how Media Cast and Media Transport worked and how future products like desktop conferencing would work. Scott found it very effective to explain to content providers how they could make money using the EIN with Application Services. Also, it was useful to show that applications could bundle the Streaming content into their apps and then make money using the Advertising, Pay per View or Subscription Revenue model.

Around this time John Hay released the edited version of the Strategic Vision[238] document which Scott had driven the development, now synchronized with the plan that had been presented to the public on Jan 20th 2000, now formally approved and available for distribution to EBS employees. Scott had worked with many people to write it, addressing the significant confusion over what the business was and what the focus would or should be.

 The Executive Summary of the Strategic Vision Document explains the concepts and ideas that Scott brought to EBS around applications riding on the EIN platform.[239] Rex Shelby, Larry Ciscon and David Berberian had brought the idea of a Wide Area Network Operating System to the table using InterAgent as the middleware messaging element. Scott had originated the idea when still at MFS and later at WorldCom but now it had a home at Enron Broadband Services and would be called the Broadband Operating System or BOS. This gave Scott a strong sense of pride, and his enthusiasm for Enron was strong.

Other Enron people, especially Tom Gros with help from Stan Hanks added such ideas as Bandwidth Trading and Intermediation.

---

[238] http://tinyurl.com/RoutingPaths-EBS-Vision
[239] http://tinyurl.com/RoutingPaths-EBS-Vision

Changing Directions, and the Dotcom Crash

It became obvious that Portland was no longer at the helm of EBS. Joe Hirko and his team were less and less in command of what EBS was about. The videos made in January, give a sense that David Cox, as Chief Commercial Officer, was driving the company commercially.

Having bought ECI, Enron Corp. was now taking over. With the new EBS christening, the checkbook moved from Portland to Houston. Clearly, EBS was to be run by Kevin Hannon (COO), David Cox (CCO) and Ken Rice (Co-CEO) but from Enron. Scott had experienced acquisitions before and knew the signs. Portland people were kidding themselves if they thought they would continue to drive the company.

In April 2000, the technology stock market crashed. Enron was an energy company that had a high-tech start-up inside of it. Most of our ex-MFS friends were all with Telecom, ISP, or DSL players. The stock market was killing high-tech startups. Particularly hard-hit were Telecom/ISP/DSL startups like my own NorthPoint Communications. Any company not EBITDA (earnings before interest, taxes, debt, and amortization) positive imploded. The crash decimated even powerful startups that raised billions like Global Crossing or Level 3 Communications.

Enron was not valued as a high-tech stock company or it would have crashed with the rest of the "tech stock bubble" players. The crash of 2000 did hurt the EBS potential customer base. EBS startup companies were planning to roll out new content and application services using the application services platform. This alone was serious enough, as it meant a reduction in the number of companies looking to buy their services.

Being a part of a larger company allowed EBS to weather the storm. Scott was excited to be part of Enron, an energy company. His friends at Level 3, Allegiance, NorthPoint and other new applications companies were hurting. Apple, Microsoft and Cisco prices crashed, but the Enron stock went up, proving Enron valuation was not linked to the EBS plan itself a dotcom play. Scott believed Enron was solid. Smart Money articles showed how Enron relatively unharmed by the crash.[240]

Scott believed this proved the market considered Enron's valuation to be based on the core energy business, not EBS. This meant being asset-light (unlike Level 3 and other carriers) was a smart position. Using pooling points to trade bandwidth meant EBS could operate well below the costs of other companies and survive the downturn. Scott was optimistic about Enron and EBS in 2000.

---

[240] http://tinyurl.com/SmartMoney4112k

Y2K: A Year of Innovation

EBS continued innovating and building on the concepts Scott brought to EBS along with Modulus, and with which Joe Hirko had secured Enron's backing. This collaborative effort was what Scott worked to keep going throughout the year. His efforts to help John Hay with training, including using the Broadband content, continued to be used. Kevin Hannon established himself as COO. David Cox established himself as the commercial driver of deals. The army of over 1,000 employees of EBS developed a robust network with sophisticated, media-rich applications, bandwidth from their own fiber, their own IP backbone, pooling points to acquire bandwidth from other networks and a Bandwidth Trading desk and application. It was all real and valuable, as many videos and supporting documents show.

Per Cox–known for successfully bootstrapping tech and trading startups–got involved at EBS and spent several weeks trying to determine its business case and commercial path forward. Innovative as Enron was, only one person at Enron had a plan, and both the plan and the person had been somehow set aside for what was thought to be more "Enron-like" opportunities like Bandwidth trading.

Cox says when he met with Yeager they decided to team up and meet with prospective customers that would potentially touch all the current EBS offerings. Scott gave Cox the Strategic Vision document that he had initially crafted for Enron communication. On their first trip to meet the late Danny Lewin of the Boston startup Akamai, Cox told Scott that he thought they should see the plan through. Scott agreed enthusiastically, but noted the new Enron management team was quickly losing faith in the plan's approach to put the application services in place and sell the applications one at a time.

Scott sketched on three pieces of grid paper what he thought would become (and did become) the modern Internet. They agreed that they had to find an anchor tenant for the Media Cast service described precisely in Scott's original plan. If successful, they knew that, as Ken Rice would say, getting a "bandwidth hog" on this network would put us on a path to success.

 Soon thereafter April Hodgeson's origination team located the "bandwidth hog" they were looking for: Blockbuster Entertainment, Inc. Cox made a historic 25-year exclusive deal with Blockbuster to deliver video to the home via the EIN. After the Blockbuster deal was announced in April 2000, Cox directed his focus to the project and turned to Scott's talents and expertise to help navigate the complex rollout of the service.[241]

---

[241] https://youtu.be/dHKzz9a6d-s

Scott had an enormous amount of experience dealing with the local-loop companies including the Regional Bell Operating Companies (RBOC). Over the next few months Scott's original hires supported the team securing Video on Demand (VOD) service distribution contracts with RBOCS that included Southwestern Bell, (AT&T) and Bell Atlantic (Verizon). Cox's excellent team, led by the brilliant Franklin Bay, set out to create a completely new infrastructure for Video on Demand for the Blockbuster deal. The Blockbuster VOD service would be rolled out in the coming months across selected U.S. markets including in Manhattan. The technology differed somewhat from the Media Cast technology in addressing a more industrial market, but captured the "bandwidth hog" required by the Yeager/Cox plan and necessary for the company to implement Scott's Strategic Vision document. Internal adversity, market resistance and coordination with the U.S. entertainment industry and 100-year-old telephone companies suggested this could not be done in a few months. But that was Enron: go big or go home. Scott attributes the unlikely success to David Cox, April Hodges, Frank Bay and the entire VOD team. He "just helped them get it started." The service was far ahead of its time.

The Blockbuster VOD service, up and running by the end of 2000, was demonstrated[242] at the 2001 Enron Analyst's Conference. Numerous[243] videos[244] showing this infrastructure and how it worked in early 2001. Within a year Cox and his team, along with Frank Bay and Steve Barth[245] were deploying the entire Video on Demand (VOD) system on a Motorola set-top box. The video clearly shows it was the equivalent of what Netflix and every RBOC in the U.S. would look like many years later.

Disenfranchised

Unfortunately, Scott's ideas and plan had been pushed aside in favor of bandwidth trading, as had most of Portland's initiatives and projects. After the Blockbuster coup, EBS leadership saw his plan they had abandoned magically reborn and in a few months, become a percentage of the market cap of the company. The new EBS leadership took credit for having put Cox into the leadership role, discounting Scott's involvement. David Cox says Scott stoically accepted the exclusion, knowing he had dreamed it, planned it and supported the entire ECI/EBS team to make it happen. The rebuff was not Scott's alone—nearly all the early Portland founders were similarly marginalized and expelled.

Scott did not attend the 2001 Conference and was not asked to help with preparation for it. He did not understand the inner workings of Enron or how they were structured.

---

[242] https://youtu.be/8Ti5TICsJvo
[243] https://youtu.be/mipplSejBSw
[244] https://youtu.be/xmI4usZJsUQ
[245] https://youtu.be/OaoR6h8S3jw

Steve Barth's 2000 training video[246] explained the War Room and how they priced deals in a structured finance way. Scott had not been used to pricing products  this way at MFS. He knew he was not considered an Enron player, but not someone they thought needed training in the Enron ways. Scott's focus and interest was on customer-driven solutions and innovation. Enron's focus was very different and not in his comfort zone.

Raw video footage shot before the AC 2000 compared with the same before the AC 2001 clearly illustrates the developing conflict: Enron people versus service provider people focused on providing quality services to customers.

ECI sales people in Portland like Jeff Foster and John McClain told Scott they were being educated on how to do things the "Enron" way, which included financial structured components. Those were different than repeatedly selling off the shelf product/services such as a T1 Circuit (MFS), long-distance minutes or Internet access. Structured financial deals were entirely custom deals and no one was including Scott in any training on how to be part of them.

Scott tried to help with projects like defining how Tiered Quality of Service, including how High QoS TCP/IP, would enable trading of IP bandwidth versus DS3 Months of TDM circuits. Stan Hanks drove this project with Tom Gros.

Fired!

Kevin Hannon and Ken Rice fired Scott the day after the Analyst Conference 2001, saying he was not cut out to be an Enron person. Scott agrees they were right. He did not join ECI to be a commercial Enron person; he never agreed to it and was never part of the deal. His talents lie elsewhere, as does his personality.

He was brought in with strategic ideas to change the industry by creating the first-ever application platform that sold applications: not bandwidth, not Internet Access, not Dial Tone, not long-distance, not hosting or any traditional Telecom services. He had done that job extraordinarily well. Scott felt fortunate to have been given the chance to bring his ideas to fruition. He knew he had done his job well and helped the company even if he had been terminated. For whatever reason, Kevin Hannon wanted him gone, but the parting of ways was professional and Scott decided to be helpful despite being fired.

The Strategic Vision document and the patent on the BOS had been filed, and a Kevin Hannon-produced PowerPoint and training document of September 2000 showed the ideas were documented, communicated and taught to the employees of EBS. This was his charge as the Strategic Development Senior VP, and what Jeff Skilling asked him to do in February 2000.

When Scott saw EBS was winding down he produced a document outlining how the bandwidth desk could go down-market and sell T1s not just DS3s via the pooling points. It was a sunk cost that would only incur incremental expense. His idea was rejected but he tried to help leverage their investment to trading of bandwidth since that seemed to be the shift from selling application services. The plan might have worked and enabled smaller customers trading at T1 speeds instead of starting at DS3 speeds.

---

[246] https://youtu.be/OaoR6h8S3jw

Scott left EBS in July 2001 and went back to work on FYI-Net with Paul and Massoud. His progress had been cut short by the stock crash of 2000 and the change in direction of key management within EBS.

When the questionable accounting practices and machinations of Enron Corporate[247] came to light, the company's stock valuation collapsed from $90.75 per share in mid-2000 to less than $1 by November 2001, prompting a $40 billion shareholder lawsuit. On December 2, 2001, Enron filed Chapter 11 bankruptcy. Many Enron executives were indicted and some were sentenced to prison. Scott was indicted and fought his case all the way to the Supreme Court, where he was acquitted, completely vindicated, and cleared of all charges. The Broadband Guys' Story as told by Scott's cousin Paul Yeager adds more detail to Scott's story.[248]

He had done the job he had been hired to do. The technology development he oversaw was real, worked, was deployed and used, fundamentally transforming the Internet. Media-rich content delivery is the norm, and millions of people watch their video on demand from sources such as Netflix, Hulu, Acorn, CBS, NBC, and CW. The DVD rental storefront characterized by Blockbuster is gone, the lone stakeholder in that space being Redbox. (Blockbuster failed to make the leap to streaming and collapsed, and Redbox attempted a streaming service and failed to make the transition, though they still hang on in the kiosk DVD-rental space.) Netflix still rents DVDs by mail, but streaming has captured increasing mindshare.

Scott did his part to drive several organizations to implement the changes behind this media-rich content environment. Without his influence and the hard work of several thousand people in sales, marketing, engineering, operations, and product development, the Internet would be a very different proposition today.

And while today's Internet may be ubiquitous, it is not safe and secure. Scott has one more revolution to lead.

---

[247] https://en.wikipedia.org/wiki/Enron_scandal
[248] http://ungagged.net/view.php?story=32

ADVANCING "COMMUNITIES OF TRUST"

When MFS Datanet, UUNET, PSINet, Suranet, and other companies enabled the first Commercial Internet via MAE-East in 1992, the first private IP service providers were focused on ubiquity and convenience. No one focused on security. No one cared about security.

Each of the key Private IP backbone players—UUNET/Alternet, PSINet, SprintNet, MCINet, and the regional SuraNet—agreed: the first priority was simply getting people to use the public Internet as opposed to the NSF-funded Internet backbone.

Beyond the need for personal password protection, no one imagined (or stated it if they did imagine it) the Internet's original "any-to-any" design would make it so easy for bad actors to hack into millions of computer systems worldwide, stealing, corrupting, spying, and sowing seeds of digital mayhem with impunity. The challenge still is not generally understood or widely discussed.

The Internet was, essentially, built to fail with regards to security. Protecting information was not even on the radar. The criteria was global connectivity of anyone to anything from anywhere with no limits. Like the early Telegraph and Teletype networks, everything—email, private chats, and financial transactions—transited the Internet in unencrypted plain text. The goal was global IP visibility and connectivity, not security or quality. The Internet was designed to allow information to be viewed by the entire network as packets of electronic information move from any point to any point. Every connection and node within the network can peer into the DNA of each packet. As a result, there is little skilled hackers can't do.

The Rise of Privacy

Irresponsible as this sounds today, in 1992 effective encryption did not exist. IPsec and TLS predecessor SSL first appeared in 1995. All the networking technologies that had existed prior to the Internet, extending all the way back to the original Morse Telegraph, sent their communications in the clear. Telephones were decidedly non-private; shared "party-lines" were a way of life in rural America for many decades. The concept of "privacy" in communications is a new concept.

The big, bold beautiful thing about the Internet was that no one owned it or controlled it, and few rules existed. From its ARPANET roots in 1969, the Internet has been shaped by structural design decisions that have had a lasting impact on our ability, or rather our *inability*, to secure our computer systems. Malicious mischief, the theft of personal data, and the threat of exploitive and disabling attacks target the very national infrastructure that we depend on as a society for our security at large.

Vulnerability to the threat of cybercrime is rampant and omnipresent. We open ourselves to the threat every time we turn on a tablet, a laptop, a cell phone, or a desktop computer. To this list, we can now also add any number of chip-activated personal and household objects known as the Internet of Things (IoT). Hackers have already demonstrated the ease with which they can seize control of these things too, from automobiles to refrigerators to security DVRs

The problems created by our technology are well known, yet inexplicably they remain largely unsolved, and we seem to just accept that things cannot be changed as though "the genie is out of the bottle."

In the "any to any" Internet no one entity owns or is in control. The perspective of a private network, however, suggests there might be another approach that could help companies, employees and end user consumers become more cyber-secure one private network at a time.

Creating an Internet Alternative with Private Networks

David Cox, a former business associate, came to Scott with an IPsec-based network encryption technology that proposed private networks be built on top of the open Internet using software. The idea led Scott to imagine how anyone might change the rules of connectivity for himself. Many companies have a finite and well-defined realm of legitimate digital interactions. Those companies and users could, with enabling software, prioritize security over global IP visibility. This idea is not for everyone or every business application but it does make sense for those that recognize a need for security.

Scott was present at the creation of rules for the Commercial Internet that ensured no one controlled it, no one could turn it off, and no one could force anyone to do anything. The Internet was not in the realm of control of any one entity. Because all routers published all routes to each other via Interexchange points for free (peering) global IP visibility was built into it.

While starting a new chapter in his professional life, Scott realized he could make up a new set of rules that companies could implement one private network at a time. Scott and Cox's idea does not propose "fixing" the Internet. It envisions a private network that may or may not use the Internet for underlying transport, and whose owners can implement rules that inspire and motivate parties to behave in a cyber-secure manner.

Scott and David knew the lack of cyber security and the ability to enforce cyber security policies across the Internet has always been a troubling problem. Many people believe the lack of effective cyber security is the world's largest single threat in 2017. President Barak Obama issued Executive Order 13636 "Improving Critical Infrastructure Cybersecurity" on February 12, 2013. Key people in the U.S. government identify the lack of proper cyber security as a serious problem.

Scott and David amused themselves with schemes "fixing" the problem. Perhaps posting rules on Wikipedia would drive people to follow them? Perhaps tweeting rules out one at a time would lead the whole world would follow them? Or maybe creating a website called RFC 1087 "How to behave on the Internet" would change everything? Eventually they admitted no simple answer will change behavior on the Internet because there is no easy solution for changing the Internet as it exists today. Scott and Cox grappled with shaping a plan that would gain viral adoption.

While at EBS Scott constantly evangelized the Communities of Interest concept, which he had identified in his original Strategic Vision document. It had been essential to the success of the EBS VOD project. Now Scott proposed to Cox a revival of the guiding principles to create a new private network called a *Community of Trust*: a new set of business and technical rules enabling private networks to become more secure one network at a time.

Its principles stemmed from the thinking espoused by the 1991 Houston Focus groups, where numerous players with different perspectives studied a variety of networks. They were based in building private LAN backbones over MFS Datanet Ethernet for Enterprise customers. MAE-East history was also brought into play. Ultimately, they considered the needs content providers wanting to stream content to a closed audience to create revenues. Scott and Cox based the Community of Trust concept on their prior experience. It would resemble the early MAE-East in that not all players had to use it, but early adopters would gain a clear competitive advantage. If it didn't catch on and become a de facto standard, it would still merit just a handful of players. Early adopters would have the advantage of the disciplines defined in the legal, insurance and technology framework of a Community of Trust.

They concluded there are indeed ways to create a new set of rules to make private networks secure. The legal framework rather than a specific technology is key. There would be ways to route information via the Internet while maintaining security and ensuring much greater privacy for end users if a new legal framework were adopted. They believed this could be accomplished chiefly through voluntary adoption of a new policy structure for Internet users that walls off voluntary participants in a way that shields them from risks without hampering their continued usage of the Internet. Scott and David agreed to found a company to take this idea forward.

I reconnected with Scott while this concept was forming, and met David to discuss his ideas. I had been exploring similar ideas while consulting with another company and had incorporated a few similar core elements into another project I was then pursuing. I instantly recognized the importance of what they were trying to accomplish and was excited to be invited to join.

A Community of Trust

We call this new private network structure a *Community of Trust* (CoT). CoTs are governed by policies that spell out how everyone is expected to behave and what happens when they don't. Each CoT will have a mechanism to "trap" any bad actors that pop up. Additionally, each CoT may use any technology to add further levels of security. The CoT is a technology-agnostic approach, focusing on standards and procedures rather than a specific technology. Nonetheless, we do champion a specific architecture. We perfected the original architecture and submitted it to Underwriters Laboratories. This system became the world's first UL certified cyber security platform.

We have begun exploring ways to help make Communities of Trust not just an option but the norm. Our company, Reprivata, is actively working to translate our ideas into practice in the real world.

Built to Fail

Before delving deeply into how a Community of Trust solution will benefit Enterprises and individuals, let's explore the origins of the problem. This is an important step to understanding why the twin problems of cyber security and cyber privacy remain unsolved. How we can provide protection via the establishment of a new structure of policies and agreements will armor our existing communications networks.

Let's begin with a simple and easy to understand summation of the problem from one of the foremost authorities in the world on Internet security and privacy threats:

> *"The Internet was not designed to resist highly untrustworthy users."*

This statement appears in bold type in a 2002 research report produced by the CERT Coordination Center in connection with the *Software Engineering Institute* (SEI) at *Carnegie Mellon University* (CMU). The research was funded by the U.S. Department of Defense. CERT stands for *"Computer Emergency Response Team."* It creator, the *Defense Advanced Research Projects* agency, also was responsible for the creation of the Internet.

CERT came into being in 1988, the year the Internet had its first major computer attack. That December, the Morris Worm appeared — a portent of things yet to come. CERT began immediately tracking the number of computer security incidents it handled. It logged six in 1988. They logged 52,658 in 2001. By 2003 the number was 137,529 and by then CERT also was receiving 542,754 email messages and more than 934 hotline calls reporting computer security incidents. The worldwide number is anybody's guess.

If the Internet was not designed to prevent security attacks, how *is* it structured? The CERT report notes: *"the original users of the Internet were university researchers, seeking to collaborate with their colleagues and openly share and exchange information for the benefit of all. ... the user community was essentially benign and trustworthy, so simple security measures (such as passwords) would be sufficient to ensure the protection of the Internet community. Under this worldview, provisions to track and trace malicious user behavior were never designed or implemented."*

Perhaps the first response to the glaring issue of security and the Internet followed the Morris Worm. In January 1989, the Internet Activities Board released RFC 1087 titled "Ethics and the Internet" to admonish those who might abuse the network, although the Morris Worm is not mentioned. A relevant excerpt reads:

The IAB strongly endorses the view of the Division Advisory Panel of the National Science Foundation Division of Network, Communications Research and Infrastructure which, in paraphrase, characterized as unethical and unacceptable any activity which purposely:

    (a) seeks to gain unauthorized access to the resources of the Internet,
    (b) disrupts the intended use of the Internet,
    (c) wastes resources (people, capacity, computer) through such actions,
    (d) destroys the integrity of computer-based information, and/or
    (e) compromises the privacy of users.

No thought other than that was given to security since "any-to-any" connectivity makes it impossible to stop anyone from doing anything. Since no one owned the Internet and no one entity could control the Internet it does not matter if more detailed rules did exist or do exist buried somewhere. No one has to follow them and the criminals, hackers, nation states or just bad actors doing what they think is cool stuff know they do not have to follow any rules anyway.

There is no central authority of any kind that can detect or enforce behavior onto people or computers run by computers run by people to do anything. In effect, it said, "be nice." It appears it was naively expected that if you published an RFC to the Internet audience they would not only listen but abide by the rules.

The focus then was getting people and business to simply use the Internet at all. It seems clear that in 1992 when MAE-East was formed, no one imagined a far-distant future of January 2017 with 5 billion devices or more interconnected.

I rather imagine that had the issue been raised to any of the Internet proponents of 1992, the response would have been along the lines of, "That would be a nice problem to have." Certainly, neither Scott nor I know of anyone that thought about what that might mean to everyone on the planet if some small percent just ignored these RFC 1087 rules and did whatever they wanted on the Internet.

And the world has ever since struggled with cyber security issues!

Internet security "incidents" almost always involve the networked communications that make the Internet possible. A "protocol suite," combining Internet Protocol (IP) with Transmission Control Protocol (TCP), technically accounts for how everyone on the Internet can "see" or "find" people and Web sources, and how we communicate back and forth — or rather, it provides the structure that allows our computing devices to do this.

The CMU associated CERT report of 2002 also noted *"despite serious security shortcomings, TCP/IP is still the standard protocol suite for network communications on the Internet, greatly limiting our ability to track and trace Internet cyber-attacks to their source."*

Information moving around the Internet is broken into pieces known as IP packets and transmitted to their destination with the help of TCP, which allows hosts to establish connections and makes sure the packets are delivered in the right sequence. The IP packets contain the pieces of information being sent, the destination address, source address, and port number, among other things. The packets travel over what is often described as "a vast array" of routers along the network, moving by any number of alternative routes, looking for the best path from any router to any router, until the packets are delivered to their destination.

It is a design that the CERT report "recognized the need to be robust in the presence of external attack," but "there was no equivalent concern with regard to the possibility of internal cyber-attacks by the Internet's own users."

I do not intend to diminish the value of the Internet. We can all enjoy the good fostered by the Internet, the innovation in communications, global collaboration, and openness. But we also now see how it's being abused.

Internet pioneer Vinton Cerf wrote a foreword to the 2013 edition of Tony Standage's "*The Victorian Internet.*"[249] His words are cautionary and chilling. "*There is a dark side that cannot be ignored. Our dependence on digital technology and digitized information also creates vulnerability. Our devices may ingest digital viruses … and other malware. Our daily lives may be disrupted. … We have created our own liabilities to go along with the extraordinary utility of our connected universe.*"

Billions of Devices

As of December 2016, there were about five *billion* devices connected to the Internet. Predictions now estimate that there will be 50 billion to 120 billion devices by 2020! This exponential growth has spawned many more targets for bad guys — and made it that much easier for them to conceal themselves.

Let's imagine for the moment that one to five percent of all people connected to the Internet are bad guys. Also, now there are nation States that sponsor cyber-attacks daily. Just imagine how many more attacks, and bad actors, we will see four years from today as the Internet continues its current growth trajectory!

What do we do? Especially since we are starting so late, and considering the complexity of the privacy and security issues we're facing? Do we shrug and say it is what it is, this is just the way the Internet will always be? Do we kiss good-bye our security and privacy?

Overcoming Inertia

The large Search Engine, Social Networking and Internet/telecom companies promote the idea that only they know how the Internet works. Only they know how to monetize and protect it. Yet they have spent almost no money developing methods that might assure users that their data and identity are secure. In fact, they have been moving in the opposite direction. And the reason is that they all have huge financial incentives to keep global IP visibility (any-to-any connectivity) just the way it is.

---

[249] https://www.amazon.com/Victorian-Internet-Remarkable-Nineteenth-line/dp/162040592X

Their mantra is always the same. Maintaining a wide-open Internet, they say, helps individuals and drives market innovations while generating sky-high returns for global free markets. Though built on several fallacious assumptions, this seems hard to argue against. The Internet environment is much less free than many believe. The Internet and the services it offers are tightly controlled in many ways. However, it is not our purpose to overturn the entrenched powers.

Rather than attack the underlying claims, we merely suggest there should be a choice. Users should have the option to use available networks in a secure manner, and the market should decide what works. Networks should be safe and responsible honest behavior the norm, and users should not be exploited unless they choose to expose themselves to the wild-west environment.

There should be a choice and until now there has been none. But that doesn't mean solutions are nonexistent. There are ways to keep information private while making it securely available to networked players, including individual consumers, Enterprise customers, employees, public sector entities, and content providers.

Information Mining

Information is valuable. In today's world, every keystroke, every mouse-movement can be monetized by "big data." The norm now is that even in a private place like the Apple content protected network your information is mined by them and others and you have no ability to block players from seeing your devices and attacking them while using a private solution like the Apple music/content network.

If you use Windows, even inside a corporate Enterprise network, Windows is constantly "calling home" to report information it has gleaned from your usage.

Most of this is innocuous, the information is merely sliced and diced, anonymized and analyzed to deduce what sells. Sometimes it becomes aggressive and annoying, as when your email provider notes your discussion with a friend about needing a roof, and then bombards you with spam from roofers. Inside a Community of Trust, this data-mining and "calling home" behavior is controlled or eliminated. Windows and macOS are unable to report back to their masters unless the Community of Trust owner and Users agree to this business policy. Spam and malware are reduced, contained or eliminated.

Taking Control

The CoT solution works for *all* the devices connected to a private network to stay connected and be private if the CoT owner so chooses.

This is a "free market" approach.

The real beauty of a CoT is that it represents a new way of thinking. It seeks to simplify matters for end users at scale while making it possible for businesses to become more secure and drive security measures back into their supply chain and interconnected entities.

Many of the legacy tech companies today act as if every end user is at least as smart as a top earning Internet security expert. As if the average computer user is going to take the time to learn the complicated and constantly changing privacy settings and security protocols to protect themselves while using, e.g. Facebook — and then spin around and learn, and execute, and stay up to date with the different settings and rules associated with Twitter or Google accounts, or the platforms associated with the constellation of social media we use.

Meanwhile, beyond the need to stay current with social media security issues, end users must also constantly attend to the ever-present flow of prompts to securely and safely make needed security "fixes" and "enhancements" that amend security gaps in software programs and apps. The constant stream of updates and downloads patching our systems is annoying and harms productivity.

Are all these changes we introduce to our operating systems truly needed? And exactly how do we weed out the Trojan horses bearing malware dressed up as a security fix? And, of course, there is the problem of phishing and having email accounts hijacked The openings for invasion are endless.

A CoT is designed to help participants navigate this crazy cyber hazard course even as it provides new levels of security and privacy.

Our simple starting point is to assume a cyber security breach *will* occur. We believe a company can reduce the likelihood of a breach by following the contract structure. Next, we define the legal rules and processes that will govern how we deal with the breaches. We believe that if we first establish a path to a solution; we set aside, at least for the moment, the challenges posed by the technology that permits cyber-attacks. Instead we first establish new rules that govern all relationships formed in cyberspace.

Only then can you quantify the risk, define the players who are interconnected and make it possible to define and limit the Community of Trust to a finite number of users, employees, and interconnected parties.

The MFS private network approach combined with the Content provider approach offers a private network delivering content to end users in a private secure manner. This allowed us to break down the problem down to a finite solution for each private network and define each as a Community of Trust.

The materials we need are readily available. The concepts are within easy reach. We see them at throughout this biography: from the technical successes behind the development of fiber optic cable communication networks and private networks to the bandwidth services applications and the initiation of the Commercial Internet itself.

Here, in brief, is how the CoT works: The CoT creates voluntary "Communities of Trust" built upon legal relationships between end users, employees, Community of Trust owners, and any interconnected third parties. Such communities may, optionally, include secure, private, encrypted software network layers that ride on top of the private network that may include using the Internet as a low-cost access approach. New technologies will almost certainly be developed to make things more secure, but the idea of a private network legally bounding the way players interact is independent of technology.

We believe CoTs are practical and workable. We can codify these business relationships in legal documents, with nothing preventing companies adopting this approach to begin defining CoTs today. It is a free market and permissionless approach like MAE-East, which provided an end-run around the NSF Government-funded network. CoTs offer a way to conduct business over a private network in a secure way that is independent of technology and repeatable across the Internet.

Finally, there is a Risk Management approach that is intrinsic to the definition of a Community of Trust. We selected the NIST Cybersecurity Framework and chose maturity level Tier 3 as the minimum level which is a repeatable level of cyber security. Then we mandated that the Community of Trust owner is CSF Tier 3, and our contracts mandate that all their interconnected third parties also achieve CSF Tier 3 themselves. Then, per the adoption of the contracts, they require all parties to obtain cyber insurance.

We use a Master Interconnection agreement for all interconnected 3rd parties that holds everyone to the same standards. This ensures the weakest link is CSF Tier 3 compliant and obtains cyber insurance. Now you have a common standard and the risk is quantifiable so that insurance companies can insure a known risk. Insurance underwriters can now write new categories of insurance based on a new ability to define, quantify and manage the risks. Real cyber insurance will be available to protect the company, the employees and the end user consumers who belong to a Community of Trust. This can happen one private network at a time and does not need to be deployed worldwide to be of an immediate benefit to the adopters.

Scott, David, and I along with others, are focused and committed to providing a viable choice to those entities that want to conduct business in a private network manner, even if they use the Internet as part of their private network solution. The idea enables those applications and behaviors conducted within a Community of Trust.

Origin of the CoT Idea

The basic practices and standards for Community of Trust concept emerged and evolved in the early days of the Internet as we introduced new ways to improve communications for large-scale Enterprises using fiber optics, local loops, and broadband services applications. We were offering, delivering and helping companies use the same basic building blocks at MFS, WorldCom, and later at EBS.

In the early days, everyone used a private network to interconnect. The Focus Group videos from 1991 in Houston explain the different approaches very clearly, those concepts gleaned from experience while building private fiber optic networks in the 1980s and a service provider network as MFS in the 1990s.

From the start, we were devising custom protocols to interconnect our devices and creating secure, private networks. The networks extended from any end user's device to the servers in use (typically inside a building or campus) called a Local Area Network, or LAN. These LANs were directly connected to a Server or the host mainframe, such as a DEC or IBM computer.

That basic, simple and well-understood structure, I believe, is what's needed to provide the companies and users who want to conduct business over networks via the global Internet connectivity with the privacy and security it now lacks.

We intend to start small, working with early adopters who are willing to design, deploy and utilize an enhanced and improved version of their private network. Later, those that want to Interconnect private networks can do so, but with a new set of business rules consistently applied. Over time this could be an alternative to the Internet, consisting of a new series of private networks for those who prioritize privacy and security. We believe the technology exists to build into the design the potential to safeguard the privacy and personal data of every end user.

I believe strongly that the creation of CoTs can assure security and privacy when they are wrapped with a legal layer, use cyber insurance, and incorporate design elements that protectively "close off" its users from the Internet.

The Internet was not something the public had access to in the early days. For the most part, a user had to be associated with an academic or research institution. Those institutions set the rules governing who could use the Internet and the purposes users would be allowed to put it to. The rules were publicly set forth in what is known as an *Acceptable Uses Policy*, or AUP.

AUPs (also known as "fair use policies") remain integral to establishing allowable practices under the information security policies that govern today's information systems, websites, and networks. In today's digitally driven information ecology, AUPs are omnipresent and written by governing bodies like businesses, universities, ISPs, and social media website owners. AUPs can be linked to other security protocols. Sanctions are set out against those who break rules and regular security compliance audits are required.

Major weaknesses to AUPs, however, make them unsuitable to helping protect privacy and provide security. They are not legally binding, nor are they supported by an enforcement mechanism. It's a "trust me" scenario for those who agree to their terms.

A means is needed to articulate and enforce how everyone on a network is expected to behave. If applied in the right way on a broad and consistent scale, such agreements can be used to enhance and strengthen security protections for everyone while neutralizing the activities of many bad actors on the Internet.

Getting the broadest array of Internet participants possible to agree to a new set of voluntary but legally binding rules that govern behavior within the unique networks I envision would change everything.

It is not reasonable to try to do this all at once. Beginning with one company and one group of users and employees, the agreement would require only of those that interconnect to that one company. Market demand should dictate adoption. This is a free-market approach and those who adopt it will be the users who value security and privacy more than convenience and global connectivity. This is an alternative approach for a finite number of players that understand the need for it, the users who value security and privacy more than convenience and global connectivity. This is an alternative approach for a finite number of players that understand the need for it. We believe banks, law firms, accounting firms, medical providers, power companies and even network providers would be able to behave in the private secure uniform manner defined in our Community of Trust contracts. The market must identify one Community of Trust at a time.

Community of Trust Agreements Structure

Recognizing today's serious threat to Enterprises and end users by bad actors using the Internet means Communities of Trust will offer much more than an AUP.

The CoT uses sophisticated contractual "master" agreements to define and structure expectations, policies, risks, protections and relationships between CoT owners and all participants within a CoT. Master membership agreements require members to implement security policies and technical specifications, for instance, and master interconnection agreements (like a financial settlement mechanism) require the interconnecting parties and the CoT owner to hold a minimum threshold level of cyber insurance. Other agreements include setting out the ways in which end users are protected, and defining the responsibilities of CoT employees.

One of the most important agreements will require the CoT service provider to be U.S. Cybersecurity Framework (CSF) Tier 3 compliant. A CoT offers a practical approach to implementing that framework and addresses U.S. Executive Order 13636, "Improving Critical Infrastructure Cybersecurity" (signed into law by President Obama in 2013) by offering a voluntary risk-based cyber security framework that is "prioritized, flexible, repeatable, performance-based, and cost-effective."

Increasingly, the public and government are looking to top board members of corporations to be held accountable for cyber security risks and plans. This means they must not only attain a level of cyber security literacy, they also need to know what governance structure to trust.

The IT Governance Institute publication "Information Security Governance: Guide for Boards of Directors and Executive Management" (2nd edition) summarizes the challenge. Terry Hancock, CEO of Easy 1 Group, is quoted: "The complexity and criticality of information security and its governance demand that it be elevated to the highest organizational levels. As a critical resource, information must be treated as any other asset essential to the survival and success of the organization."

When the Internet Went Public

When Rick Adams and UUNET embraced the proposition that MFS could provide a Metropolitan Ethernet that allowed ISPs and their routers to interconnect in the Washington D.C. area, we were collectively creating a unique, independent network — one no longer riding on the National Science Foundation backbone. Once we provided Rick Adams with a national Ethernet backbone that also connected to MAE-East (one of four critical national Internet hubs), he gained a private network that was his alone, with its own routers. We set in place a new architecture supporting the rapid emergence of the commercial backbone of the Internet in 1992-93.

Other *Internet Service Providers* also started building regional backbones. Some, like PSI Network Inc., built national backbones and launched migration from private networks to the "public" Internet we have today. Of course, once we had service providers connecting to each other and creating "inter-networking" through MAE-East, we needed business and technical rules set forth. The most important of these stated simply that any router could and would send traffic to any other router, no permission needed. No single entity could dictate, or in any way "control," the path taken by information as it moved from a starting point to an ending point. In fact, no one would need to "decide" anything. The protocol we adopted would make things happen automatically.

Any end user connected to, let's call it, ISP #1 could "talk" to any end user or server connected to ISP #2 through MAE-East, and so forth. The importance of this any-to-any movement of data packets over the Internet cannot be overestimated.

The idea first arose because of a simple and straightforward conversation with Rick Adams.

"Well, who's going to own this thing?" he asked. "We don't want one person in charge. And we don't want one player who didn't pay their bills to bring the whole thing down."

Nonpayment was a concern to many at the time. Al Fenn, then president of Datanet, on being told that Scott was planning to sell Ethernet services to ISPs. "They don't pay their bills," he said. "It'll never happen. It's not commercially viable." "But," Scott replied, "we can turn them off if they don't pay their bills."

Scott heard the concerns of Rick Adams and the other members of MAE-East. Together they devised the specific rules that allowed this new way of connecting and sending traffic from any given service provider to another without permission from anybody, including the federal government.

We intentionally kept the business side simple: you paid your connection fee and agreed that the whole thing could not, and would not, be shut down if any ISP went belly up, failed to pay its bills, or even turned off its routers, causing that connection link on the Internet to go dark. It was imperative that operations and information move freely, without interruption. No one entity could control or stop this peering from happening. Once there were multiple Layer 2 peering points (i.e., Ethernet/FDDI) then no single point of failure existed and this permissionless approach to running the Internet was in place.

These same rules operate across the Internet today and are critical to an understanding of the privacy issues under discussion.

The "any-to-any" rule also created conditions allowing anyone connected to the Commercial Internet to "see" anyone else, a wonderful achievement promoting global connectivity. You can stick Web servers anywhere and connect them to Internet servers. A Web server is no different than an end user, so you can see those, too. Later, when DNS, or "domain name service" was developed, the "any-to-any" functionality we'd created ensured that all the domains on the World Wide Web would be visible to anyone.

The business and technical rules we implemented over MAE-East — which mandated all ISPs could see each other through DNS, as well as at the IP layer — became the catalyst for the Internet's growth and the opening of the network to the general public's use. We'd created a network that was permissionless, operated based on global IP visibility, and was literally wide open. Anybody could connect and send anybody any data they wished. There was no authority to stop them.

New types of businesses flourished. And a new economy was born. That has been the chief reason for Internet businesses to flourish and why some new idea can go viral and attract new audiences overnight: when permission is not required, information spreads of its own accord. Rick Adams and early ISPs like PSINet and others imagined the Internet being a utility like electricity available to everyone.

The business and technical rules upon which the Commercial Internet rests have been the force behind the creation of the Internet as we know it. Facebook, Google, Apple, Microsoft, Amazon, Twitter, Snapchat and other sites have been the force behind the creation of a free and open Internet.

In 1993 the University of Illinois, Urbana-Champaign came out with the first freely available Web browser. Mosaic gained a million users quickly and helped to fuel the Internet explosion. In 1994, two students at Stanford University, Larry Page and Sergey Brin, began developing a search engine they took commercial and named Google. Google created a search engine, gave it and other applications like email and maps away for free, and it was adopted globally by over a billion people without anyone asking permissions or controls. Facebook and Twitter have followed suit.

In the following 10 years, the number of Internet users exploded at lightning speed: from one million to 10 million. And then it leapt from 100 million to a billion. Experts say that now we're approaching, or perhaps surpassing, two billion connected devices.

This underpins the new global economy we see expanding at an accelerated rate with worldwide impact. Let's consider for a moment what this means. Given the near global inability to protect privacy on the Internet, in an interconnected world in which everyone can see everyone else, a world in which all our devices are "visible" and thus open to attack, what happens as the number of devices expands?

Let's agree to identify the widespread problem of identity and data theft as the work of individuals or groups we'll go on calling "bad actors." In today's wired world, bad actors can range from nation-states to an individual with a laptop sitting in a car outside our favorite "free W-Fi" coffee shop.

And let's say the number of people connected to the Internet who are bad actors represent some arbitrary percentage. The figure could be as low as 1 percent of all Internet users. Now, what happens as the number of users grows from one billion to 10 billion? The percentage of bad actors grows in a corresponding way. Suddenly we have 10 million, or 20 million, people willing to become bad actors because they can "see" two billion other people.

A single individual today can access servers that allow him to "see" the equivalent of thousands, if not millions, of people, and can use those servers to mount an attack. The bad actor's reach is even greater when multiplied by the number of nodes globally that connecting us all to the Internet.

The Internet provides ubiquitous connectivity for the dissemination of free information and the development of free Enterprise. It also can become a dark and dangerous space. We are all enabled equally by the very thing design put in place: "any-to-any" connectivity.

A lot of people have no idea that the Internet as we know it was constructed in this way on a rule promulgated by a handful of people in 1992 through MAE–East. Its challenges, however, are in the forefront of general awareness today, as terms like "cyber security" have entered the vocabulary of every owner of a computing device. Each new theft of information carried out by ghostly and seemingly uncatchable cybercriminals and other "bad actors" reminds us of the threat we now collectively face.

People naively and incorrectly assume privacy on the Internet. Consequently, among recent developments in the consumer goods arena, the ability to turn off and on "location" service functions is pre-installed on cell phones. We have a limited authority to prevent others from physically tracking us through our devices. By installing encryption programs, manufacturers like Apple seek to reassure consumers certain identifying information can't routinely be accessed by the manufacturer, a telecom company, or a government agency. Tech engineers say privacy is being "baked" into our devices.

Social media companies are likewise scrambling to shore up the perception that they take user privacy concerns seriously. In 2016 Facebook made big waves with its announcement that it would notify users when it believed they had been targeted by hacker attacks suspected of originating in hostile nation-states. Directing the user to Facebook's "login approvals" setting, protection consists of setting up a typical two-factor authorization process. The user then receives a text from Facebook with a login code whenever someone tries to access the user's account.

What the Internet Reveals About Us

The wealth of information available about any of us who visit Facebook is astonishing, and it may be used by a hostile government, a local hacker, the U.S. government, or even a friendly advertiser. One study based on a simple analysis of Facebook "likes" demonstrated "the degree to which relatively basic digital records of human behavior can be used to automatically and accurately estimate a wide range of personal attributes that people would typically assume to be private."[250] [Proceedings of the National Academy of Sciences V.110 n.15] The analysis was based on "a dataset of over 58,000 volunteers who provided their Facebook Likes, detailed demographic profiles, and the results of several psychometric tests." The simple model used to produce the results ran subsets of "likes" data through logistic/ linear regression computations and out popped predictive individual "psycho-demographic" profiles.

From these "easily accessible" records of "likes," Cambridge University researchers could quickly and accurately glean the "sexual orientation, ethnicity, religious and political views, personality traits, intelligence, happiness, use of addictive substances, parental separation, age, and gender" of the people who volunteered to take part in the study. Facebook suddenly appears less a place for casual exchange of news and information with friends than a space where, with each login, mouse click, and "like" we unwittingly write revealing autobiographies.

On some social media websites, the personal information ready for invasion may already be neatly packaged and ready for hackers to exploit. No linear regression required. Certainly, this was the case for the 37 million users of AshleyMadison.com, a website catering to married people looking for extramarital hookups. Hackers bypassed the website's security protections with seeming ease, taking names, addresses, and credit card information, and then publicly exposed the identities of the website's users.

Sadly, such tales of massive security breaches are now daily affairs. The motivation behind them is as diverse as the Internet itself. The intent may be robbery, social embarrassment, or the acquisition of information on U. S. citizens, foreign governments, corporations, and political candidates. The same information used to empty your bank account can help someone gain influence and power, enable them to subjugate and blackmail; to propagate garden variety mischief, cause a company to lose millions of dollars, or even humble a national government.

A hack in November 2014 of Sony Pictures and Entertainment involved the public release of sensitive business information. The hackers demanded Sony pull its pending distribution the film *The Interview*, presumably for its unflattering depiction of Korean leader Kim Jong-un.

---

[250] Proceedings of the National Academy of Sciences V.110 n.15
http://www.pnas.org/content/110/15/5802.full

Big data breaches are far-reaching. JP Morgan Chase has reported 76 million records stolen, as have Home Depot (56 million), Zappos (24 million), Evernote (50 million), Adobe (36 million), eBay (145 million), Target (76 million), Living Social (50 million), Experian/T-Mobile (15 million), Anthem (80 million), Hannaford Brothers (4.2 million), Heartland (130 million), JetBlue, Dow Jones, Euronet (2 million), as well as British Airways , the European Central Bank, Dominos Pizzas France, Apple, Japan Airlines, Yahoo Japan, Twitter, Snapchat, NASDAQ, Neiman Marcus, Uber, UPS, Twitch.tv, Vodafone, 7-Eleven, JC Penny, Visa, Jordan, Rock You!, AT&T, AOL, Citigroup, TJ Maxx, and Ameritrade.

Government hacks, meanwhile, roll right along with their own drumbeat of media attention. At one end are information dumps of hacked secrets exposing government activities to the public, including the precise ways in which the U.S. government spies on citizens. Some information has come from former CIA employee and privacy activist Edward Snowden. Others originated from WikiLeaks, the international publisher of secret information from anonymous sources founded by Julian Assange. We have the example of 50 million Americans who lost their personal data when hackers managed to worm their way into the U.S. Office of Personnel Management. Similar serious big data breaches involve the IRS, U.S. Department of Veteran Affairs, and the U.S. Military, as well as hacks of White House data. To this growing list, we could add universities, hospitals, county governments, and additional government offices. Even Secretary of State Hillary Clinton had a private secure server that seems to have been hacked. The server was on the Internet, visible globally to anyone.

As a test, Reprivata put a new server on the Internet. In six minutes, it was attacked by more than 10 groups of bad actors from all over the world, despite having no content or information.

This endlessly growing list of entities hacked by bad actors, and the financial and social costs to individual users of the Internet showcases obvious Internet risks But these do not stop people from using it. In 2017 Facebook users numbered 1.79 billion, and that number continues to grow.

Economy, infrastructure, and daily lives around the world are now dependent on computers. So how do we protect ourselves against the theft of information and who we see, where we go, what we do, where we keep our money and how we spend it?

An August 2015 article in *US News*, "The Illusion of Online Privacy," asserted — correctly, I believe — that the Ashley Madison Hack only highlighted the reality that "Web companies can't guarantee privacy." The article provided a shocking perspective on security breaches that occurred in the previous six months. As 2015-16, about 505 data breaches had targeted businesses, government agencies, and other institutions, exposing more than 139 million records, per the Identity Theft Resource Center.

The sheer number of hacks is evidence of how companies underestimate the threat of a data breach and how the government needs to procure software faster to keep up with the latest cyber-security technology, according to Bruce Schneier, a fellow at Harvard University's Berkman Center for Internet and Society.

The difficulty in protecting data is straightforward. "Everything that touches a computer produces data – and your data moves around a lot," Schneier explains.[251] "There is not much you can do to protect it because you are not holding your data. We are relying on other people who hold that data."

The solution to the problem of privacy should recognize that the Internet is very much like the old-fashioned newspaper that landed on the home doorstep. Internet companies depend, as newspapers and magazines did (and still do) on advertising and thus on sales of everything from books to boutique wines and toilet paper— only now it's done online. The difference, of course, is in the structure and inherent power of the Internet to allow Web companies like Google and Facebook to track every purchase, preference, and interest. In exchange, we get "free" services, such as email and searches. How do we take back control of our personal data and at the same time promote the Internet as this amazing place of interconnectivity and commercial freedom?

With major breaches occurring weekly, this is no longer an academic exercise. I just read a report that said while 30-40 percent of big Enterprises are claiming to have been the subject of cyber-attacks, a private study of the healthcare industry found the number is probably closer to 95 percent.[252] In June 2015, CBS Money Watch reported more than 80 percent of U.S. Companies have been successfully hacked.[253] Small Business Trends in April 2016 reported that attacks are increasingly targeting small businesses, with 43 percent of recent attacks directly aimed at small business.[254]

---

[251] http://www.usnews.com/news/articles/2015/08/25/the-illusion-of-online-privacy
[252] http://resources.infosecinstitute.com/category/healthcare-information-security/healthcare-attack-statistics-and-case-studies/ - gref
[253] http://www.cbsnews.com/news/percentage-of-companies-that-report-systems-hacked/
[254] https://smallbiztrends.com/2016/04/cyber-attacks-target-small-business.html

No Immunization for Companies

It's not to everybody's benefit to report breaches. In fact, now a company can be held liable under criminal and civil law, and potentially put out of business, if the company attacked self-reports the attack and someone thus decides to file legal action because obviously, the company didn't protect them against known risks.

There's no immunization law to help companies that are attacked. They can be sued by consumers. They can be sued by their vendors. They can be sued by their employees.

That was another huge consideration as we worked through our new CoT approach, which we think of as an inevitable structure. Inevitable, in the sense that people are ready to migrate to something that's safer and more secure than the wide-open Internet. CoTs provide safe harbors for all parties — except bad actors.

> *I can't just call up Google and put in an online order for all the information the company has collected about me. I can order virtually any product on earth via online shopping. I can use a search engine to research the most obscure facts about things and people — and get instantaneous results. But government and the marketplace have yet to provide the public with a simple and direct way to learn what information Internet companies hold in their electronic bowels about us.*

We took a broad look at existing problems today and sought a comprehensive solution. We looked at all the bad actor attacks. We looked at the fact that Google and Facebook and all the other Internet giants own our digital identities. We looked at how every search, Facebook visit, and online purchase generated data about each of us collected and maintained in databases that we have no control over.

Not one database, mind you, but multiple databases are capturing data about us at the same time. If I go to a website, my activities are being tracked through at least four or five, and perhaps as many as 10 or 20 different cookies. Ten or 20 companies own all that data; I don't own any of it.

And if I want to go and get that data about myself and look at it because I'm curious about what the Internet knows about me? I can't do it.

I might get some of it — if I'm creative, aggressive, and persistent. But I can't just call up Google and put in an online order for all the information the company has collected about me.

Yet I can order virtually any product on earth via online shopping with a click. I can use a search engine to research the most obscure facts about things and people and get instantaneous results, again with a simple keystroke. But government and the marketplace have yet to give us a way to learn what information Internet companies hold in their electronic bowels about us.

We don't know what these companies are doing with our digital identities. And there appears to be no limit to how deeply they may pry into our personal online activities.

Internet experts will disagree on many things, but they will generally agree on this: nothing is private on the Internet. If you want privacy you must create it. And to create an effective privacy shield you must be skilled and conversant in many areas of computer science and its offshoot disciplines. Everyday end users, even those highly sensitive to the privacy issue, lack the basic understanding to do anything about it.

These thoughts contributed to the impetus behind our drive to create the CoT idea and envisioning a different future for our networks. But this was only part of the overall picture.

The bad guys can get to you from anywhere using the nature of the Internet to do it. That's called the attack surface, and it is very large.

Then you have the digital identity of the end users being captured by all kinds of companies, as already noted, and being monetized. So, that was another public concern we set out to solve.

Multiple Issues, Single Solution

These all might seem like separate issues. Cyber-attacks involved more issues than end users' privacy and ownership of digital identity. And they're different than the issues involved in advertising fraud or piracy of content. At least that's the way people tend to think about these things, right?

But as we looked at these things we said, "Wait a minute. Really, if we did it right, we could solve each of these problems with one solution.

We realized that if we created a community of trust, and that community of trust used software to encrypt all traffic from end-user devices — a computer, or smartphone, an iPad, or whatever — then that traffic would no longer be visible on the Internet.

The "any-to-any" visibility of global IP addresses would end. The DNS ability to go to the world-wide web of servers while inside the community of trust would not be permitted. We would abruptly seal off a major attack surface that villains target. We would also put a new set of business and legal rules in place to limit bad actors' activities and would track activities in a new way.

Safeguarding Users Data and Digital Identity

We used the NSA Mobility Specification to create a software package following those rules. This package uses the strongest encryption algorithms recommended by the NSA. We realized that if we created a Community of Trust around this software along with the legal and insurance framework, it could be secure. All traffic from end-user devices — a computer, or smartphone, an iPad, or whatever — would be encrypted using secure, private encryption certificates and assigned a private IP address to these devices. Then it would be encrypted again by the application-layer traffic provided by that outer encrypted tunnel. That traffic would no longer be visible on the Internet.

The "any-to-any" visibility of global IP addresses would not exist inside this encrypted Community of Trust or Encrypted Core. Malicious operators on the Internet could not peek inside the tunnel. Malicious behavior inside the tunnel would be blocked and contained. Connectivity to World Wide Web destinations while inside the community of trust is eliminated. Major attack surfaces targeted malicious operators are sealed off. There is no possibility of infecting the user through malware-infested web pages. Would-be villains are unable to access resources cloaked inside the Encrypted Core. We would also put a new set of business, legal and technical rules in place to limit bad actors' activities. The technological element allows us to track nefarious activities in a rigid and controlled manner.

Simply put, a technological layer is added to the legal and insurance framework. Placing encrypted content and applications inside the Encrypted Core provides double protection. Common applications such as email may be encrypted using custom, private certificates.

There is a second layer of encryption for VoIP and XMPP, so phone calls and text messages also would be encrypted.

Even if an attacker should penetrate the Community of Trust encrypted core, security would remain intact. There is no navigation permitted to unauthorized destinations, even with the IP address. A stolen certificate, for example, only permits access to authorized resources. The Encrypted Core blocks access to all resources not expressly permitted. We have immense latitude in managing users and blocking attempts to break the rules.

The rules make the difference. The CoT exists within a new set of legal rules. When creating the Commercial Internet through MAE-East, we created a set of business and technical rules that governed connections and disconnections, and determined traffic would flow to everybody. Those rules within MAE-East essentially dictated there would be no rules. But the rule within the Community of Trust is that there are many rules. When you join a CoT you're forced to live by those rules. And what are you going to get for agreeing to the rules? Your digital identity is re-privatized, and you're in control of it.

That data is going to be in a data warehouse in an encrypted database as secure as the traffic that flows across the network.

The data will be anonymized. If you did permit a search engine inside the CoT to import information from the wide-open Internet, this would occur in a secure manner. Managed access to the external data source would be via white-listing, and tight routing controls. You would not even be individually identified. You'd appear as a generic searcher and reveal only the information specified in your CoT contracts. You'd control whether an Internet vendor could, for example, learn your age, or your geographic location, or your personal preferences. And you would participate in any financial gain from the sharing of that information.

The financial rules govern the Community of Trust and the relationship between content sources and end users. End users get to control their digital identities, which are stored in a secure database called the "Central Privacy Authority." The privacy authority entity is the steward of the trust and all information about its members. But even it can't look at the data inside of the privacy authority. It's just a warehouse.

Once you have real people in a Community of Trust, and an assurance of real end users versus an unknown quantity of bad actors in place, advertisers can be assured that they would have real eyeballs and real people viewing and clicking on their ads. Any botnets within the Community of Trust would be detected and reported to the Community of Trust owner using several possible technologies. The Reprivata UL Certified CoT software would make it very difficult for malware to gain entrance to the Community of Trust, and would also capture unauthorized traffic to help catch bad actors.

Advertising Protection

David Cox has spent years fighting advertising fraud. This experience and knowledge proved invaluable in shaping the Community of Trust solution.

David determined first-hand that massive advertising fraud existed by representing content creators. He wrote software that determined where on the Internet their stolen content resided. This led him to prove that advertising networks and advertisers were unwittingly wasting advertising dollars by paying for sites that stole their content. The bad guys were making money from stolen advertisements and botnets simulating non-existent eyeballs.

AdWeek reports this is a $7 billion problem annually in documented cases alone.[255] Digital advertising sites like Excelaca are exhorting the industry to act. Recently, new research uncovered an elaborate cyber crime ring with a new sophisticated botnet known as Methbot, which could "view" some 300 million video ads a day, generating up to $5 million in stolen daily revenue. Forbes described Methbot as the largest ad fraud network ever.[256]

Cox developed ad-tracking software to detect the physical location of viewing. His approach proved effective in addition to remaining hidden by the entity stealing the content and creating false clicks.

The advertising dollars are flowing to the bad guys so they are active and innovative. The same bad guy that will steal your digital identity also has a website of stolen content. And that same bad guy creates bots and false names to attract false eyeballs to his site of stolen content. They get paid real dollars for those faked or misrepresented eyeballs. They have the incentive to devise innovative methods to steal the content and identities with which they attack sites on the Internet and defraud advertisers.

People think their identity is not worth stealing. They ignore the breaches and other problems. There is a far greater problem than most people imagine. Malevolent entities are making billions of dollars from leveraging global IP accessibility. Cox sees this first hand in the form of billions of bits of false data within the huge ad fraud problem.

Advertisers think that they are paying for real impressions. In 2014, the advertising business was worth about $70 billion North America alone. Reports say 30 to 50 percent of all online ad dollars is money paid to sites claiming bogus impressions. The advertiser paid for the ad impression without receiving the information they wanted.

Cox discovered many methods used to defraud advertisers: ways to to steal content, display false advertisements, create revenues from real and fake eyeballs. He recognized the need to detect these issues in real-time. This, with knowledge of the ubiquitous connectivity problem, led him to develop a solution in the form of a software platform.

---

[255] http://www.adweek.com/news/advertising-branding/whats-being-done-rein-7-billion-ad-fraud-169743

[256] http://www.forbes.com/sites/thomasbrewster/2016/12/20/methbot-biggest-ad-fraud-busted/ - 25e026b64ca8

Cox learned that digital advertising creates a dangerous cyber-attack surface. He discovered most Chief Information Security Officers are unaware of this exposure. They underestimate the risks created by a company's digital advertising operations. Criminal abuse of digital advertising networks creates more than just fraudulent transactions. Criminal redirection, site cloning, and spear phishing are among the many cyber threats. Criminals mount these and many other exploits via insecure and unaudited advertising networks.

He developed the Brand Protection Controller API, which eliminates the opportunity for criminals to access customer-facing networks. It creates a secure audit trail by inserting a beacon into the advertisements. Tracking this beacon assures advertising network compliance with the Brand Protection Controller instruction. With this, he created the Advertising Compliance Authority (ACA) Ad Network API. This eliminates ad fraud by requiring the ad network to request the ACA to serve an ad on a domain or in an application. The ACA must approve each domain/app before it can register impressions. The ACA will approve valid requests for ad insertion in real time. Requests from a robot or criminal infringer generating fraudulent impressions or clicks are denied.

Cox knew how to manage the enormous volume of data in the digital ad ecosystem, and created a unified real-time stream processing engine: a high-performing, fault-tolerant, scalable, asynchronous in-memory platform capable of processing billions of events per second. It is also fault-tolerant. It can recover from node outages with zero data loss and without human intervention.

Cox and Scott know how sophisticated the bad actors on the Internet are. Knowledge and experience help us recognize how fast they morph and find new ways to attack. Experience enables recognition of the role of "any to any" connectivity of the Internet.

Cox's insight combined with Scott's background created a unique perspective. We realized that we needed something like a Community of Trust legal/insurance framework. This must happen by driving secure human behavior into the Enterprise, one private network at a time. Those entities that do adopt this approach will benefit immediately, independent of global adoption.

Reprivata

A Community of Trust is safer because the "Wild West" wide-open Internet vulnerability has been removed. Its legal framework helps define the business relationships. Once the Enterprise chooses voluntarily without asking permission from anyone to use the legal framework, a new set of business rules drives the Enterprise to behave in a more cyber-secure manner.

We called the company Reprivata because the Latin "res private" root means private property, "private way" or various connotations of privacy depending on context.[257] A Community of Trust is a private network but also it connotes the idea of reprivatization of the business rules of network regarding Enterprise.

A CoT is like a private space: it is a private network. A CoT also re-privatizes, or gives back, a person's digital identity in the way privacy is handled in the Central Privacy Authority. Scott approached Andy Lipman, ex-MFS Chief Regulatory officer, with the idea of reprivatization of company networks. He wanted to help them to make up rules for gaining control of their private networks, and require everyone they interconnect with to play by their new rules as well. This puts control in the hands of the owner of the private network, who can deal legally and technically with issues one private network at a time. Andy agreed this concept was entirely valid in the legal regulatory world. By making them private, the Interconnection agreements could be made up and defined by Reprivata, empowering the Enterprise that adopted these rules and standards to be deployed and forced back into their own ecosystem. Lipman—arguably the worlds most experienced and knowledgeable interconnection attorney—felt the idea was legally valid and that to his knowledge was not being contemplated by any other carrier or service provider. Scott was assured of the concept's validity, but also of its originality. Time and money could safely be invested in developing these concepts.

The CoT Master Agreements drive the Owner of the Community of Trust to achieve CSF Tier 3 status, after which every interconnected entity must also achieve CSF Tier 3 status. All parties must purchase cyber insurance based on this legal framework and achieving CSF Tier 3 status. This makes the Community of Trust much safer because it changes the behavior of the players and it brings the weakest link up to the standards which are achievable and affordable.

Once a Community of Trust Owner chooses to use the Master Agreements they have the choice of deploying the technology themselves for no additional cost other than servers and hosting, or they can use a Community of Trust Service Provider to provide the UL Certified multi-layered encryption technology as a service.

The use of the software to build a closed network or a software Community of Trust overlay means users cannot see everybody else from within a community of trust. Similarly, the servers within the Community of Trust are not visible on the wide-open Internet.

---

[257] https://en.wikipedia.org/wiki/Res_publica

We created the concept of the Central Privacy Authority because we believe each Community of Trust needs to have "warehousing data in a secure manner for the private network" spelled out legally independent of technology. This concept is explained in all the contracts. It serves to define and protect the privacy of the end users and the employees. It defines the role of the Community of Trust Owner and the methods by which all interconnected parties might share information for cyber security purposes. This assists in meeting the CSF Tier 3 requirement, especially when collaborating with 3rd parties. It also helps flag and address bad behavior within the private network.

The entire Internet ecosystem that exists today, from vendors to software companies and various services, can easily become part of this Community of Trust and Central Privacy Authority ecosystem. We believe the competitive market of existing cyber vendors and services are part of existing networks, which helps Community of Trust justify the expense of requiring CSF Tier 3 compliance over ad hoc deployment of existing technologies.

Community of Trust Service Providers — Lots of Players and Innovation

The movement to CoTs will require a lot of players, a lot of innovation, and a lot of capabilities. Even a billion-dollar company couldn't do it alone. With a central privacy authority supervision, Enterprise customers, content providers, advertisers, consumers, insurance companies, medical companies, and anyone else who might join could make their own versions of Communities of Trust — but they would all use the same standard set of contracts.

That's the main thing we're bringing to the table. A standard set of contracts spell out all these relationships in writing, vetted by multiple types of players in the industry. A custom version created for various industries is probable, but the standard master service agreement and all the standard sub-agreements will apply across any industry and across any community of trust, as well as to any vendor.

IBM, Hewlett Packard or Google are welcome to do a CoT, but they must all play by the same set of new rules. They just must subject themselves to the contracts, agree to privatizing data owned by end users and to end-user control of their data, and accept the technical and business rules and standards.

Those standards deal with software implementation and technologies and how one subjects oneself to audits. Presently, no one subjects themselves to audits in any meaningful or effective manner. Internet companies seldom permit a third party to tell them how good or bad they are at being secure and private.

The recent Yahoo breach revelations are instructive. In 2013 Yahoo was known to be using an obsolete and deprecated encryption mechanism, MD5, but there was no mechanism to mandate an upgrade. In the new world of CoTs, CSF Tier 3 audits become standard, with compliance required by the contracts. A routine CSF Tier 3 audit would reveal the twenty-year-old broken hash and the offender would be required to upgrade.

By taking steps like these we create accountability in the Communities of Trust that does not exist on the Internet today. Holding the end user accountable to stated behavior restores their identity. It also holds companies and the Community of Trust accountable to delivering services the way they claim they will. And it's holding the third-party vendors that interconnect to be accountable in playing by the same set of rules.

Employees of the Community of Trust also are held accountable to stipulated rules while protecting them and others. The Community of Trust owner will be required to meet the CSF Tier 3 specifications and obtain cyber Insurance.

These rules will be standardized across all Communities of Trust. When the Central Privacy Authority requires these rules of all the members and everyone agrees to play by them, the end-user is protected.

When the UL Certified, multi-layered software is deployed it will help keep the bad actors out. They won't be able to easily steal content or get to any resources from the Internet. Even if they become Community of Trust members, they can't see or harm other users. Everyone who joins legally subjects themselves to rules which include the ability of the Community of Trust Owner to look for bad behavior and stop it if. People in these communities are protected and secure. The networks are reliable and robust.

I've been working with networking for many years now. Scott and I started in the 1980s with packet switching, local loops and local networks and home offices. Then in 1991-92 the Internet arrived and we found ourselves present and involved as MAE-East gave birth to the Commercial Internet.

David Cox deployed the first commercial Video on Demand network and has innovated with respect to content protection and advertising fraud for the past eight years. He developed the software that I worked to document and obtain the UL 2900 CAP certification for as the first independently tested (UL) commercially available open source software. Many stories about his commercial and technical experiences stand behind the Reprivata Communities of Trust ideas.

And now, today, some 24 years later, we are all still working with the networking challenges posed by the Internet. We all believe that Communities of Trust will be deployed one private network at a time and therefore the CoT model is the future.

COMPANIES AND ORGANIZATIONS

VERY IMPORTANT PEOPLE

TECHNOLOGY AND INNOVATION

REGULATORY AND FINANCIAL

www.ingramcontent.com/pod-product-compliance
Lightning Source LLC
Chambersburg PA
CBHW081309170526
45166CB00011B/3456